Understanding the Jet Stream

Clash of the Titans

Ged Dunkel

authorHOUSE®

AuthorHouse™ UK Ltd.
500 Avebury Boulevard
Central Milton Keynes, MK9 2BE
www.authorhouse.co.uk
Phone: 08001974150

© 2010 Ged Dunkel. All right reserved.

No part of this book may be reproduced, stored in a retrieval system, or transmitted by any means without the written permission of the author.

First published by AuthorHouse 5/5/2010.

ISBN: 978-1-4520-0732-8 (sc)

This book is printed on acid-free paper.

Contents

Contents 1

Chapter 1 — An Introduction 1
   1.1 Working with Longitude and Latitude 4

Chapter 2 The Jet Stream — A Brief History 10

Chapter 3 — What is the Jet Stream? 12

Chapter 4 — Air Masses 18

Chapter 5 — The Coriolis Effect 21
   5.1 Introduction 21
   5.2 The Coriolis Effect 21
   5.3 An Everyday Example 22
   5.4 A Large scale Example 23
   5.5 Conclusion 25

Chapter 6 – High and Low Pressure Systems 27
   6.1 High Pressure Systems 29
   6.2 Low Pressure Systems 30
   6.3 Convergence and Divergence 30
   6.4 The Characteristic Rotations of Highs and Lows ... 31
   6.5 The Measurement of Atmospheric Pressure 34

Chapter 7 - What are Weather Fronts? 37
   7.1 The Cold Front 39
   7.2 The Warm Front 42
   7.3 The Occluded Front (Occlusions) 44
   7.4 The Stationary Front 51
   7.5 Frontal Lows (Depressions) - Formation 52

Chapter 8 - Common Jet Stream Profiles Affecting the UK 66
   8.1 Your Starter for 10 67

Understanding the Jet Stream – Clash of the Titans

    8.2 The Atlantic Jet 70
    8.3 The Northerly Jet 72
    8.4 The Southerly Jet 73
    8.5 The Arch Jet 75
    8.6 The Horseshoe Jet 76
    8.7 The No Jet 77
    8.8 Summary 79

Chapter 9 – An Introduction to the Weather Charts 82
    9.1 Low Pressures 83
    9.2 High Pressures 84
    9.3 The Fronts 91
    9.4 Isobars 92

Chapter 10 – Forecasting using JS and Atlantic Charts 96
    10.1 Electromagnetic Radiation (EMR) 98
    10.2 Satellite Imagery 100
    10.3 Forecasting 105
    10.3a Cold Front 106
    10.3b Warm Front 107
    10.3c Occlusions 108
    10.3d Passage of a Frontal Low 109

Chapter 11 – The Endish 115
    11.1 – Practice Makes Perfect 115
    11.2 – Can You See the Jet Stream? 116
    11.3 - The Jet Stream – Can Be Quite Nasty 116
    11.4 - The Jet Stream and Climate Change 117

Chapter 12 – Basic Cloud Types 122

Chapter 13 – Useful Web Sites 124

Chapter 14 – Glossary 126

Chapter 15 – Acknowledgements 128

Chapter 16 – The Front Cover Explained 129

Chapter 1 — An Introduction

This book is simply an interpretation of my observational notes made over the last year or so and is intended as an introduction to the workings and effects of the Jet Stream in the United Kingdom. In writing this, I am trying to talk to you and not lecture; some sections will get a little heavy, but I do try to lighten them up a little.

In the UK we hear very little about the Jet Stream through the media only the occasional fleeting mention, when the weather is bad and the weatherman tries to appease the public by blaming it on the Jet Stream. However, I do believe that, in the USA, the weather forecasts often contain details about the Jet Stream, its position, strength, and effects.

Kirsty McCabe, of the BBC Weather Centre, recently said during an interview: "I think in the UK we are slightly different in that we perhaps don't credit the viewer with enough understanding of the weather, and I think that, you know, perhaps we will just have to start slipping them in gradually".

Well, let us hope that it will start soon. But it is not really worthwhile unless the public knows how to interpret Jet Stream information. I have tried to find non-technical books on the subject, but with no luck and that is the reason for this little book.

Over the past few months I have been trying to explain the Jet Stream to my Mum (who is 88 years old), and now, instead of commenting to her friends "what an awful week: rain, wind, cold and it's the middle of July!", she now tells everybody that "it's that horrible Jet Stream again", and she's proud of it!

Human nature seems to accept bad news if we understand the reasons for it, the weather being no exception. Take two extreme examples:- the exceptionally hot July/August of 2006 and the bitterly cold January/February of 2009. Both were a result of the Jet Stream not being where it should have been at that time, I will explain further in later chapters.

Of course, we never grumble if we get mild weather in the middle of winter (January 2007).

This book is comprised of two distinct components: Chapters 1 to 7 cover a bit of theory, and the remainder puts the theory into practice. I would have liked to make it all practical, but, without some understanding of how and why the weather behaves as it does, it would not make a lot of sense. If you are familiar with the ins and outs of the weather, you may still enjoy reading Part 1 - even if just for the odd chuckle.

Please note that the use of ***italic bold*** means that there is an entry in the glossary. The majority of theory only applies to the ***mid-latitudes***, where the weather is fickle and changes from day to day; this causes havoc for the weather forecaster. For a few years, we lived in Kuwait, just inside the ***subtropics*** (29° north), where the weather forecast was oh, so simple: March to October, hot and sunny; November to February, mild and sunny! Throw in a couple of thunderstorms and the odd sandstorm, and that was it - at least you always knew what to wear!

Chapter 1 – Introduction

If you are unfamiliar or do not fully understand how **_longitude_** and **_latitude_** work, have a look at the next section, 1.1, where you will find them explained; otherwise skip it and go to Chapter 2.

Understanding the Jet Stream – Clash of the Titans

1.1 Working with Longitude and Latitude

Remember that a circle can be divided into 360 degrees, and that is longitude and latitude. Unfortunately, I wish it were that simple. Just a bit of imagination is needed for the world map (Figure 1.1); it represents the surface of the earth straightened and flattened out into two dimensions (up/down and left/right). The two lines are "randomly" placed to act as reference points, the vertical line we call the Greenwich meridian and the horizontal line the Equator (we could have called them anything, Fred and Charlie; it doesn't really matter). We can now pinpoint any position on the "earth" by knowing its distance from the Greenwich meridian and from the equator. But there is a problem: if we just said the position was 100 miles from the equator and 100 miles from the meridian, this would be ambiguous. Which way - left or right, up or down? This is overcome by stating which way they are: above the equator is north, below is south and left of the meridian is west and right is east – clever, eh?

Chapter 1 – Introduction

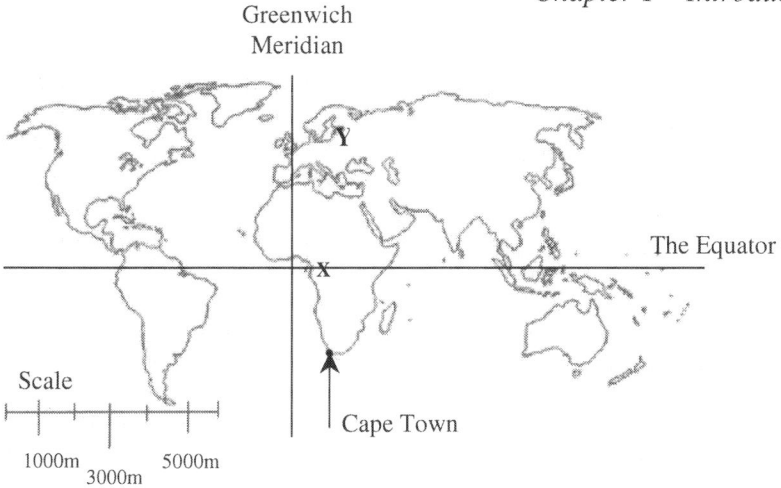

Figure 1.1

For instance, the position of Cape Town is about 1200 miles east and 2400 miles south, and London is zero miles east (it is on the meridian) and 4000 miles north. So, the unique coordinates for Cape Town would be 1200 E, 2400 S and those for London would be 0 E, 4000 N. Please remember, these figures are purely for illustration and have no bearing on reality.

Now let's make it more realistic. The earth is not flat, it is a globe, and I want you to understand why we cannot use a measurement like the mile. If you were to walk, on the equator 1000 miles to the east you would end at the position X above. But, if you walked 1000 miles due east from London, you would be somewhere around the "foot" of the Y. What has happened? The X and Y do not line up; from London you are standing "further" east than you would at the

Understanding the Jet Stream – Clash of the Titans

equator. You see that we cannot use this as a method for pinpointing positions on the surface of the earth.

The reason for this oddity is that the further north you are, the smaller the circle around the earth – and just to emphasise how silly this measurement would be, consider that if you were to stand about twelve miles inside the Arctic Circle, and walked the 1000 miles due east, you would end up back where you started from!

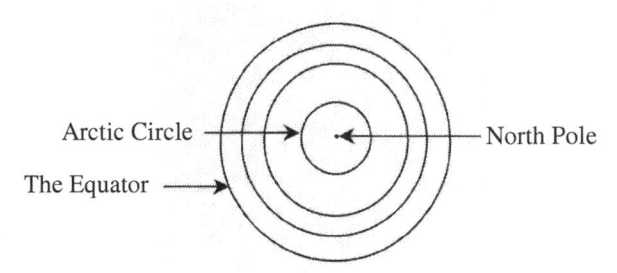

Figure 1.2

So how do we solve this problem? The problem arises because the earth is a sphere, so, intuitively, we should use the properties of circles. Say, for example, you were to walk ten degrees east at the equator you would be near a town called Libreville, in the Gabon. At ten degrees east from London, you would be just south of Hanover. Check this on a world atlas and you will see that these towns are lined up - Hanover is directly above Libreville. More on longitude later.

The concentric circles, Figure 1.2, are called lines of latitude and they are usually drawn in steps of ten degrees. Figure 1.3 shows how a line of latitude is measured. The angle shown is about 30°, and it is

Chapter 1 – Introduction

above the equator, so all the points on that line of latitude will have a latitude of 30° north.

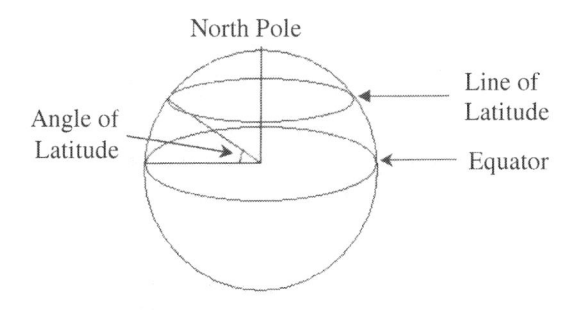

Figure 1.3

In terms of the earth, a degree of latitude is quite large, but, luckily, degrees are subdivided into sixty minutes (symbol: ') and each minute is divided into sixty seconds (symbol: "), so you might see something like 52° 56' 59" N. This is the latitude of a Met Office weather station at Watnall, Nottingham. The equator has a latitude of 0°, the North Pole 90° north and the Arctic Circle 66° 19' 48" N. Not only can latitude be written as degrees north or south, but also as positive or negative – positive is north of the equator, and negative is south.

Come on, not too bad so far.

If latitude is a measure of the "height" above or below the equator, then what about the distance from the Greenwich meridian? This distance is known as longitude.

Take a look at the two diagrams below. Figure 1.4 shows a beach ball, well, it is supposed to be the earth, with some imaginary lines

7

drawn around the circumference, each of which passes through both the North and the South Pole (these are examples of what are know as Great Circles). Figure 1.5 shows the same beach ball viewed from above the North Pole. The dark line on both diagrams represents the Greenwich meridian (notice it only extends over half the circle in Figure 1.5). This is the starting point, and this line is said to be 0° longitude.

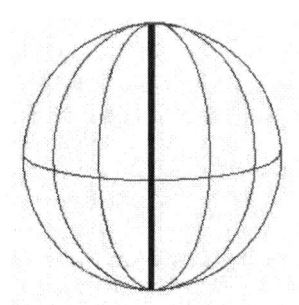

Figure 1.4

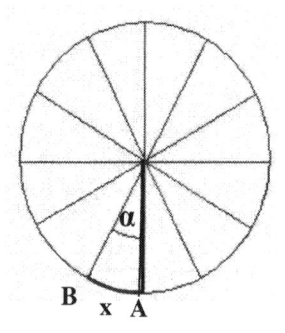

Figure 1.5

Figure 1.5, Point B is to the west of A, and instead of using the distance **x**, which we know can lead to all sorts of problems, we use the angle **α**. In Figure 1.5, I have only shown lines of longitude

Chapter 1 – Introduction

spaced at 30° intervals, so, if point B is on the equator, it has a latitude of 0° and a longitude of 30° west.

As with latitude, there are conventions: east of Greenwich is positive, and west is negative. The weather station at Watnall has a longitude of −1.1481°, which tells us that it is 1° 08′ 53″ west of Greenwich. This, combined with its latitude of, 52° 56′ 59″ north, precisely pinpoints the position, which is unique and no other place in the world has these exact coordinates. Job done!

Often you will see longitude and latitude written in decimal format: this is so that they can be used in calculations.

Also notice that the dark line (the Greenwich meridian) in Figure 1.5 does not extend from the centre (the North Pole) upwards; this is because the meridian starts from the North Pole, passes through Greenwich (London), crosses the equator, and ends at the South Pole. But this is only half a circle, what has happened to the other half? The Greenwich meridian is defined as longitude 0° but when we follow this line through the South pole it is now "on the other side" of the world and now has a longitude of 180° and so is no longer the Greenwich meridian. This means that longitudes can only be in the range -180° to +180°.

So, going back to the earlier example, the coordinates of Libreville are latitude 00° 23′ N, longitude 09° 27′ E, and Hanover's are latitude 52° 28′ N, longitude: 09° 41′ E – the longitudes are not identical, but they are close enough to illustrate the principle. Libreville is not quite on the equator, and Hanover is not at exactly the same latitude as London (51° 31′ N), but again, close enough.

Chapter 2 The Jet Stream — A Brief History

It is not known how long the Jet Stream has been in existence in its present form, but it's probably many thousands of years (since the end of the last glacial period — about 10,000 years ago). So, the history is not very brief, but it was not until the 1920s that a Japanese meteorologist, Wasaburo Ooishi, first documented it. He was interested in monitoring high atmospheric winds using pibals (**pi**lot **bal**loons) and actually documented very high-speed winds at the upper limit of the troposphere. His work was not published outside Japan, and so he remained largely unknown.

The expression "Jet Stream" is credited to a German meteorologist, H. Seilkopf, who, in 1939 wrote about the "Strahlströmung".

Worldwide acceptance of the existence of the Jet Stream came in the 1940s. Surprisingly, it was not confirmed by meteorologists but by USA bomber pilots.

Part of America's strategy to outwit the Japanese was to develop a high altitude bomber capable of precision bombing. The specification was that the plane should be capable of flying above 30,000 feet, so it would be out of range of ground flak and Japanese fighter planes. But, to be able to accurately bomb from this height, required the use of new radar technology – the result was the creation of the famous B-29 Superfortress bomber.

After months of successful trials, the time had to come to put it all into practice. While flying sorties from air force bases in the Pacific islands to Japan, the bomber crew frequently found that it was

Chapter 2 The Jet Stream - A Brief History

impossible to lock the bombsights onto the target – the bombs always overshot the target. On one of these occasions the pilot requested confirmation of their ground speed and was told by his radar operator that they had a 125*knot* (144mph) tail wind. "Impossible!" was the cry.

This frequently happened and, eventually, people accepted that this phenomenon was caused by the Jet Stream.

It is also reported that a ground speed check was requested on a reconnaissance plane's return flight (flying east to west), minus 2 knots was the reply, i.e., the plane was flying backwards!

Since the advent of the commercial jet age and long-haul flights, the Jet Stream also has an impact on trans-Atlantic flights. When flying, say from the UK to the States, the pilot tries to avoid the Jet Stream, and in doing so might detour many miles, using a lot more fuel and significantly increasing the journey time. Conversely, though, on the return flight the pilot will try to "hitch" a ride in the Jet Stream, reducing fuel consumption and cutting the flight time by up to an hour!

Chapter 3 — What is the Jet Stream?

The most accepted definition of a Jet Stream is "a river of **westerly**, high-speed, high-altitude wind circling the globe" Well, that does not sound particularly interesting, but you will see what a dramatic effect it has on our weather. I always say that the Jet Stream not only influences our weather but, can also, dictate it.

The title of this chapter is "What is the Jet Stream?"; this implies that there is only one Jet Stream – wrong, there are four. Two are in the northern hemisphere and two in the southern hemisphere.

Here I will get a little technical (but not for too long) – in meteorological terms, the earth's atmosphere is divided into six lateral zones and six vertical zones.

The Lateral Zones:-

Name	Latitude Range
North Polar	60°–90° N
North Subtropical	30°–60° N
North Tropical	0°–30° N
South Tropical	0°–30° S
South Subtropical	30°–60° S
South Polar	60°–90° S

Table 3.1

You might ask "Why are the Northern and Southern Tropical zones not considered as one, as they are contiguous (i.e. from 30° south through the equator to 30° north)?" Meteorologically, they are

Chapter 3 – What is the Jet Stream

separate because of the way that the air circulates away from the equator. Firstly, in weather terms, the equator is not an infinitely thin line but has a width of about 1200 miles. Secondly, the air over the equator is naturally always very warm and moist, and so it is continually rising. It continues to rise until it reaches the tropopause (see Table 3.2) and starts to spread out like a blanket towards the poles.

Why does the rising air stop at the tropopause? Simply, we all know that the air temperature decreases with height (approximately 6°C per 1000 metres, for warm, moist air – see Chapter 7 – The Cold Front); this applies to the troposphere. However, in the stratosphere, the temperature actually rises (approximately 2.4°C per 1000 metres); this is mainly due to the reaction of sunlight on the abundant ozone. The tropopause is the zone where this change over occurs.

For air to rise (by convection), it must be warmer than the surrounding air. But, for instance, when huge storm clouds rapidly grow and soar skywards, they enter the tropopause, which acts as a lid on the rising air, because it is as warm as the cloud's air. Hence, they abruptly stop rising and start spreading outwards. Strange, you may say, but it is a phenomenon that most of us have witnessed – the massive anvils formed above the storm clouds (cumulonimbus); the flat top is where the cloud has hit the tropopause.

Understanding the Jet Stream – Clash of the Titans

The Vertical Zones:

Name	Approximate Height in km
Outer space	>95
Mesopause	87–95
Mesosphere	50–87
Stratopause	43–50
Stratosphere	20–43
Tropopause	13–20
Troposphere	0–13*

Table 3.2

*The height of the troposphere varies with latitude and the seasons. Thirteen kilometres is the average height for mid-latitudes.

All our weather occurs in the troposphere – so there's no reason to worry too much about the upper zones; however, the Jet Streams are sandwiched just below the tropopause.

The scene is nearly set: the inter-lateral zones form permanent regions of alternating high and low pressures, due to the regions of predominately warm rising air (creating low pressure) and sinking cooler air (creating high pressure) – see Chapter 6.3.

Therefore, starting at the North Pole, this is an area of cold sinking air (so a high pressure area), we have a low pressure region between the North Polar and the North Subtropical (60° north), a high at 30° north, a low at the equator, take a look at Figure 3.1 below. Then there is a high at 30° south, a low at 60° south and a high at the South Pole. As we all know, when there are air pressure differences,

Chapter 3 – What is the Jet Stream

the high pressure tries to move towards the low, in order to even out the pressure differences.

For further reading: start with the Hadley Cells (see Figure 3.1 This is the name given to the global air circulation between the equator and 30° north and south) and Ferrel Cells (also see Figure 3.1 - the global air circulation between 30° and 60° north and south).

We just about have it; we must not think small, like a car tyre and punctures, but globally - where massive amounts of air are involved at different temperatures and pressures converging at these inter-lateral zones. It is obviously a boiling pot for something strange. *The Clash of the Titans* is where the major air masses (tropic, sub-tropic and polar) "collide" - this is where we find the Jet Streams.

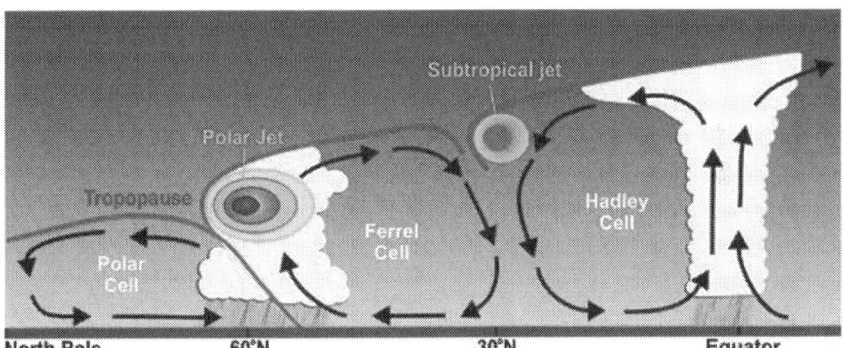

Kind permission of NOAA's National Weather Service
Figure 3.1 The Clash of the Titans

Remember, the latitudes of the inter-lateral zone boundaries are not set in stone. As with the equator, they have widths and are not fixed; they are approximations, to help explain the general global air circulations.

Understanding the Jet Stream – Clash of the Titans

Just to finish off, the naming of these Jet Streams (not difficult) starting from the most northerly and working down, are the North Polar Jet, North Subtropical Jet, South Subtropical Jet, and the South Polar Jet.

This précis is very simplified as there are many other factors involved (rotation of the earth, friction, the Coriolis effect (see Chapter 5), temperature gradients between the air masses, etc.).

The complete mechanics of the Jet Streams are still not fully understood. Research is still being undertaken, and I suspect it will be a long time before we completely understand the workings of the Jet Streams.

Below is a diagram (Figure 3.2) illustrating the position and shapes of the Jet Streams – this is one of the better drawings, as it does not show the Jet Streams as *sinusoidal* waves, but, rather, irregular.

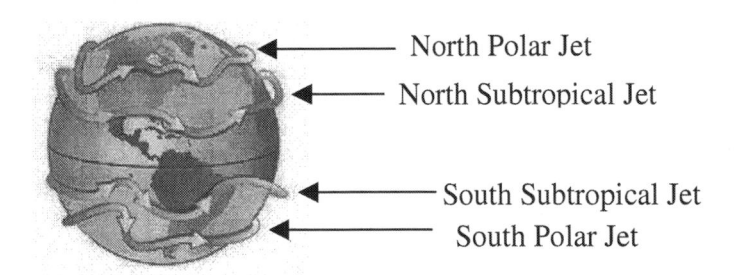

Kind permission of NOAA's National Weather Service
Figure 3.2

Unfortunately, it does show them as continuous (or apparently so) usually they are fragmented.

Chapter 3 – What is the Jet Stream

They can be as straight as a ruler, loop back on themselves, or even fade away. Their widths can vary from a few hundred kilometres to over two thousand, and the thickness can vary from one to four kilometres.

For a high altitude wind to be classed as a Jet Stream, its speed must be in excess of 60 knots and can be up to 200 knots. This maximum is arbitrary, though; in the Outer Hebrides, in 1967, a speed of 350 knots was recorded!

From here on, I am only going to discuss the North Polar Jet Stream, as this is the one that greatly influences our weather. It is also important because it affects the major populated areas, shipping, and air traffic routes.

Chapter 4 — Air Masses

It is important to understand the concept of air masses. Firstly we must think large scale (in meteorology terms **synoptic** and **global**) - thousands of miles. An air mass really is large scale, covering many thousands of square miles in area and extending upwards to the top of the troposphere. You may say to yourself "Surely that describes all the air surrounding our planet"; well, yes, but for it to be called an air mass, it must display certain characteristics. Now you can say, "I knew there would be a *but*". If, at any particular height, throughout the whole area, it has the same temperature and water vapour content, then it is said to be an air mass.

How on earth can this happen? Sounds impossible! To achieve this sort of uniformity, the air must be stationary for a reasonable length of time (many days to weeks) over a surface, at a stable temperature and humidity. Under these circumstances, the air mass obtains its uniformity of temperature and humidity from the surface it is covering.

Stationary air can only occur where there is a large area of high pressure (low pressure systems tend to be very transient); the reasons for this are beyond the scope of this book.

So, when and where do these conditions arise? It is a function of the global air circulation patterns and seasons. Below are two diagrams showing the areas of high pressure during winter and summer.

Chapter 4 – Air Masses

We see from these two diagrams, and, based on the above conditions, the areas where we would expect to find air masses are the subtropics in winter and summer, and over Eurasia during the winter. Omitted from these maps are the Arctic and Antarctic polar regions, as described in Chapter 6, as they are usually regions of high pressure.

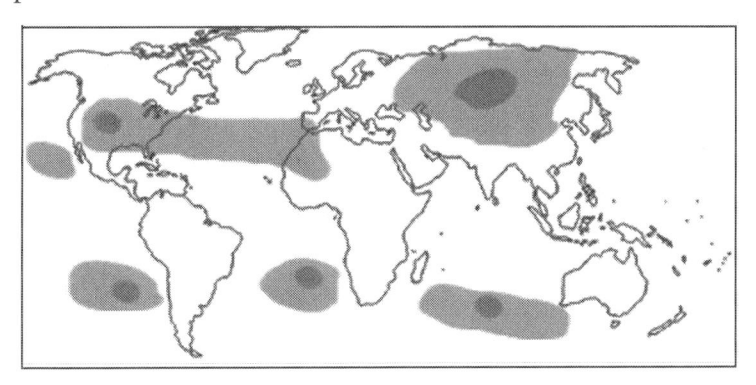

Figure 4.1 – The Winter Distribution of High Pressure Areas

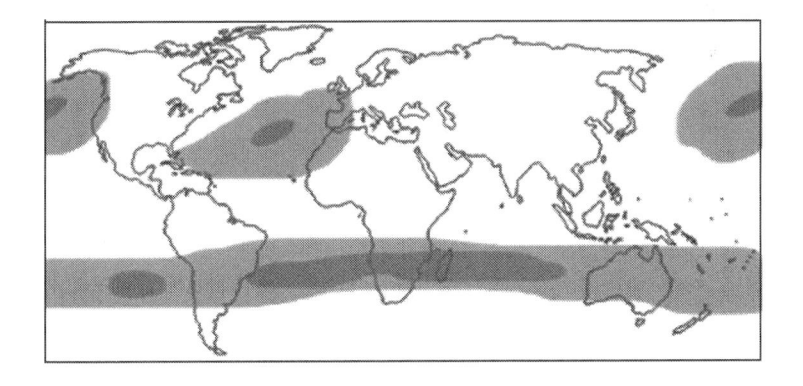

Figure 4.2 – The Summer Distribution of High Pressure Areas

The properties of the air masses can now be determined by knowing where the air mass originated:-

Understanding the Jet Stream – Clash of the Titans

Type of Surface	Region	Air Mass Properties
Ocean	Subtropics	Warm and Moist
Ocean	Polar	Cold and Damp
Land	Subtropics	Warm and Dry
Land	Eurasia (winter only)	Cold and Dry
Land	Polar	Cold and Dry

Table 4.1

Upper atmosphere wind forecasts are a good indicator of the origin and type of air we can expect over the UK. The Jet Stream forecasts, that I use, show, not only the Jet Stream, but also lower speed winds giving an overall picture of the air source (see Chapter 8).

For instance, if the Jet Stream arriving over the UK has travelled over the Atlantic from the Caribbean, then the air is going to be moist and warm. If, on the other hand, it originated from the Arctic then it is going to be dry and cold.

One can assume that the air, on its long journey, will get modified (losing/gaining heat, moisture and pollutants) by the surface conditions that it passes. An air mass reaching the UK from the Caribbean has to travel over cooler waters, so, inevitably, it will lose some of its heat and perhaps gain moisture.

The Jet Stream may give us an indication of the ground-level wind direction if the Jet Stream is directly over us (see Chapter 8) the Atlantic, northerly and southerly jets. But for the other profiles, the local ground-level wind direction and speed is governed by the local high and low pressure systems. The temperature, moisture, and purity depends entirely on the source of the air, which really can only be determined, again, by the local high and low pressure systems.

Chapter 5 — The Coriolis Effect

5.1 Introduction

You will be pleased to know that this is nearly the end of the technical stuff, and you might ask yourself, "Why do I need to know all this? I'm only interested in the Jet Stream not taking a degree in meteorology" The reason is that in order to complete the story of the Jet Stream, it is important to have a basic understanding of the fundamentals. They are all related: the Coriolis effect influences global air motion, the Jet Stream is responsible for frontal lows, the Jet Stream can only develop on a front, and so on.

5.2 The Coriolis Effect

Oh dear, this sounds as though it is going to be hard work, can't even pronounce the title! It was named after a French scientist, Gaspard-Gustave Coriolis, who first documented it in 1835. Unfortunately, to have a good understanding of how everything fits in with the Jet Stream, being aware of the Coriolis effect is a necessary evil. A lot of people struggle to come to grips with the Coriolis effect so I will try to give a simple explanation. But, before we go any further, let me just clear up a common misconception. Often you will read or hear about the Coriolis Force (later, I occasionally use this term, but only because it sounds better in context) but it is *not* a force (see below, 5.3). It is an effect caused by the rotation of the earth (or any other rotating frame), as you will see, I hope.

5.3 An Everyday Example

Imagine you are in a children's playground with an old-fashioned roundabout, and a short distance away there is a wastepaper bin. You stand on the edge of the stationary roundabout facing the bin, at X in Figure 5.1, and you throw a ball straight into the bin easy, you should be in the cricket team.

Now lets make it a bit more difficult. You go to the bin, pick up the ball, and stand where you were before. Someone comes along and starts to spin the roundabout in an anticlockwise direction, and, before you get dizzy, you have to again throw the ball into the bin. So, you wait until you are in line with the bin, and you throw it, with the same force as before, towards the bin (Figure 5.2). Lo and behold - you miss by a mile (you're fired from the cricket team)!

So, what went wrong? You've guessed it, because you were standing on a rotating frame, you should have taken into account the Coriolis effect!

Why did this happen? When you first threw the ball it only had two components to its motion, the speed at which it left your hand and the direction towards the bin.

Chapter 5 – The Coriolis Effect

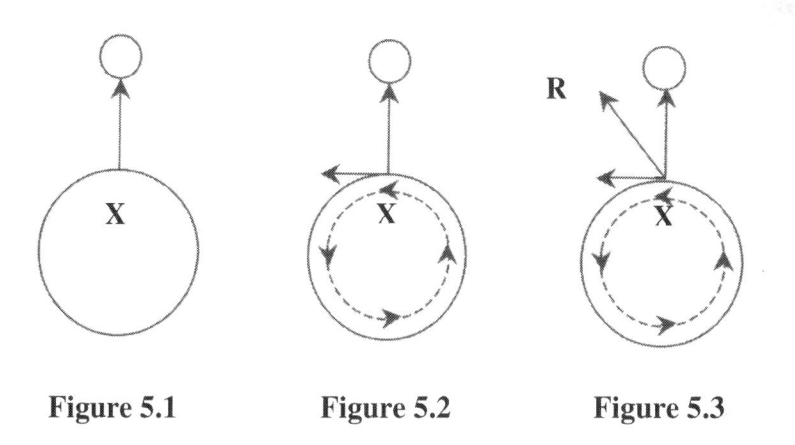

Figure 5.1   **Figure 5.2**   **Figure 5.3**

Your second attempt is shown in Figure 5.2. Not only does the ball possess the speed that you gave it towards the bin, but also, because you were moving with the roundabout, and unbeknownst to you, it was also propelled to the left, this resulted in the speed and direction shown by the arrow R in Figure 5.3. No wonder you missed! This explains why it was originally thought there was a mysterious force pushing the ball to the left (as in this example, people were not aware of it); hence it was called the Coriolis *Force*.

5.4 A Large scale Example

That wasn't too bad, was it? So, let's have a look at another imaginary example more in context with the weather (because of the great distances involved). Now we are going to use the earth as the rotating frame. You are sitting in the middle of the Sahara Desert, miles away from anywhere, at longitude 0° and latitude 25° north. You are going to fire a missile at a large target, London (longitude

Understanding the Jet Stream – Clash of the Titans

0° and latitude 51.5° north), but you are going to ignore the Coriolis effect, so you will fire it due north. Let's see what would happen (don't worry too much about the maths).

The equatorial radius of the earth is 3,990 miles, and, assuming the earth is spherical, then the circumference at 25° north is 22,720 miles and at 51.5° north it is 15,607 miles. So, if you are standing in London, then in twenty-four hours you travel eastwards 15,607 miles (650 miles per hour), but sat in the Sahara you would travel 22,720 miles in a day (947 mile per hour). This is now getting interesting!

The distance between London and where you are sitting is 1,844 miles. Say the speed of the missile is 2,000 miles per hour then it will take just over 55 minutes (0.922 hours) for it to reach London. So, you program the missile to cut its engines after fitly five minutes. But, in fifty-five minutes, London will have moved eastward 599 miles, and your site in the Sahara will have moved 843 miles. Assuming no air resistance, then your missile will also have moved eastward by 843 miles. Oh dear! You missed the target by 274 miles to the east – your missile fell on a small town in the Netherlands called Nijmegen.

You may read something like this "Coriolis effect, in the northern hemisphere, will always cause objects to veer to the right" When I first read this, my immediate reaction was, "No that can't be correct". What happens, for instance, if I fire the missile to the south? Surely it will veer to the left. Let's have a quick look and see

Chapter 5 – The Coriolis Effect

what would happen. When you fire the missile southward, towards the equator, you are moving to the east slower than you would at a site nearer the equator. Let's put some figures in, for a location directly south of you, on the equator: 0° north and 0° longitude. Then, as above, you are travelling at 947 miles per hour but at the equator the eastward motion is 1,045 miles per hour. The missile would again miss its target, but this time, to the west - because it would always be behind. After an hour, for instance, it would be 102 miles to the west of the target. But, in this case, in the northern hemisphere, yes, I was correct it veered to the left! What was really meant by that statement was the veering to the left or right only applies when you are facing the direction you fired your missile. In this case, facing south, your missile has again veered to the right.

5.5 Conclusion

Well, it wasn't too difficult, was it?

In the roundabout example, you were, in actual fact, standing on two rotating frames, the roundabout and the earth, but, because the distance between you and the bin was relatively quite small, the Coriolis effect due to the earth's rotation was negligible.

The Coriolis Force affects all fluids and not just air. There is a fallacy about water going down a bath's plughole being spun by the Coriolis effect, but, in truth, the speed, distances involved and the time it takes to empty the bath are not sufficient for the Coriolis effect to have any influence on the flow of the water.

Understanding the Jet Stream – Clash of the Titans

In further sections (see 6.4) you will see how the Coriolis effect greatly influences high and low pressure systems, as it is responsible for the characteristic rotation of these systems.

That's about it, except to say that the maths, in the real world, is quite complex. Other factors have to be taken into account, e.g., the air resistance and any original east/west components of the original motion of the object.

All that is important though, is that you understand what the causes and effects are of the Coriolis Force.

There are a lot of good Web sites that give more information, with some excellent animations. Have a look at:

http://ww2010.atmos.uiuc.edu/(Gh)/guides/mtr/pw/crls.rxml

And the absolute last word, I promise, it is food for thought. The concept of different frames moving relative to each other is the basis for Einstein's Theory of General Relativity. Bet you never thought you would read something like that in a book about the weather!

Chapter 6 – High and Low Pressure Systems

I think we should have a closer look at high and low pressure systems. I have briefly referred to these in earlier chapters.

So, what do we mean by *pressure*? In this context we are talking about air pressure. Air is a gas, which is a mixture of many different types of molecules, including mainly nitrogen (78.08%) and oxygen (20.95%), variable amounts of water vapour and trace amounts of other elements. These individual components, however small, have weight, and, collectively the whole of the earth's atmosphere, believe it or not, weighs $5x10^{15}$ tonnes ($4.9x10^{15}$ imperial tons)! For those not familiar with the nomenclature it is a five with fifteen zeros after it, so it is 5,000,000,000,000,000 tonnes, and, working the sums out, it means that the average weight is 14.7 pounds per square inch or 1.03 kilograms per square centimetre.

The population of the earth is about $7x10^9$, so that is just over seven million tonnes per person. It's a good job we are not taxed on it - but you never know what the future will bring!

Anyway, that is air pressure; it's just the weight of the air above us. Now that we know what pressure is, what is the rest of this chapter about? Well, this average global weight would only apply if we lived on a planet that did not rotate and the sun did not shine. But, thankfully, we live on earth, and we do rotate on our axis, every twenty-four hours. The sun warms the air and the ground, but the downside is that these factors complicate the pressure that the atmosphere exerts on the earth's surface.

The earth's rotation and warming cause air to move vertically (up and down) and horizontally, and this affects the amount of air above a particular point. If the air is rising, then there is less air above; this is called a low pressure. If the air is sinking, then there is more air above; this is called a high pressure. That is more or less it, but don't let this simplicity fool you; the air movement and, consequently the highs and lows, are responsible for all our weather, good or bad.

Before we go on, let me clear up an area that can be very confusing. You will hear a weatherman say "a low pressure is moving in", while another one will say "a depression is moving in". So, the synonyms for highs and lows are:

A low pressure system (often just called a low) is also called a *cyclone* or a *depression*. When the pressure of a low is decreasing, it is said to be *deepening*, and if the pressure of the low is rising, it is said to be *filling*.

A high pressure system (often just called a high) is also called an anticyclone, and, when the pressure of a high is increasing, it is said to be *building or intensifying*; also, if the pressure of a high is falling, it is said to be *weakening*.

You will read later that, in the northern hemisphere, lows always rotate anticlockwise and highs always rotate clockwise. I remember

Chapter 6 – High and Low Pressure Systems

which direction they rotate, by thinking that a high usually brings good weather (see below) and clockwise is a "nice" direction. Because lows are the complete opposite of highs, it follows that they therefore rotate anticlockwise.

6.1 High Pressure Systems

A high pressure area is defined as an area where the air pressure is higher than the surrounding air. This definition can be somewhat confusing, because logic implies that it cannot be an area, but a single point, to satisfy "higher than the surrounding air" then all the air surrounding it must be at a lower pressure; hence, it must be a point. This is true, but, thankfully, they do not just appear out of the blue; there is a large area (it can be many thousands of square miles) surrounding the point, so that the pressure gradually increases towards the centre.

High pressure systems develop where the air is falling (see Figure 3.1, Polar and Subtropic Highs, where the air is falling).

In section 7.1 you will read that rising warm, moist air cools as it rises and so is apt to create clouds and rain, and the opposite is also true: falling air warms as it descends, which makes it unlikely for clouds and rain to form. Therefore, high pressure systems are associated with settled, dry weather, tend to be slow moving, and can persist for many days or weeks.

6.2 Low Pressure Systems

Low pressure systems are the complete opposite of highs.

A low pressure area as an area where the air pressure is lower than the surrounding air. Hence, there is an area surrounding the centre of the low where the pressure decreases as you move towards it. The area of this surrounding air is usually much smaller than that associated with a high. Yes, you've guessed - they usually develop as a result of rising air, which is prone to the formation of clouds and hence the possibility of rain. But, thankfully they are usually fast moving and can come and go within hours.

6.3 Convergence and Divergence

These are just names given to the movement of air in and out of high and low pressure systems. As the air rises within a low, at ground level, this creates an instantaneous vacuum, and so air is drawn in to replace it - the air is *converging* into the low. The rising air proceeds until it eventually "spills" out of the top of the low; the air here is said to be *diverging*.

Again, the complete opposite happens in a high. The falling air is converging at the top of the high and descends to ground level where it "spills" out and so diverges.

Chapter 6 – High and Low Pressure Systems

6.4 The Characteristic Rotations of Highs and Lows

To reiterate, in the northern hemisphere highs rotate clockwise and lows rotate anticlockwise, but why? I think the following will explain.

Figure 6.1 below is a diagram representing an ***air packet*** leaving a high pressure region (northern hemisphere) and how it progresses towards a low pressure area. The scale is synoptic, the distance between the High and Low perhaps a thousand miles.

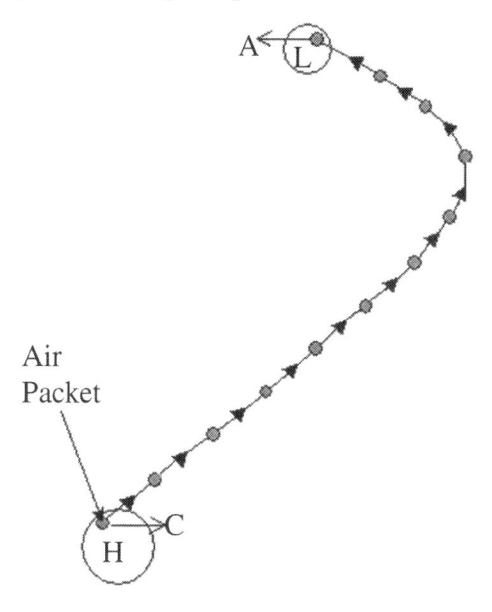

Figure 6.1 - The Northern Hemisphere

Notice that the air packet moving outwards (diverging) through the high pressure is affected by the Coriolis Force, in that it is veering to the right (this gives a clockwise component to it's motion, arrow C)

31

Understanding the Jet Stream – Clash of the Titans

– this is the reason that high pressure systems always rotate clockwise.

As the air packet moves northwards, it is drawn towards the low pressure (convergence), so it starts moving towards the low. The low is moving slower than the air packet, due to the Coriolis effect, so it is overshooting its "target" and hence starts to move towards the west. As it enters the low (convergence) it is veering to the left (this gives it a anticlockwise component to it's motion, arrow A), causing an anticlockwise rotation.

When you first look at this, you might think, "But, surely, if the air packet leaves at the bottom of the high, its movement is anticlockwise". Remember, in 5.4, if you fired the missile south, it would deflect to the west; this is true here, so it will still move in a clockwise direction (see Figure 6.2).

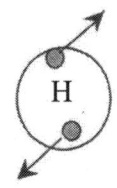

Figure 6.2

This has all been a bit heavy-going, so time for a quick Limerick:

> What was that about the Coriolis Effect
> That to the right it will deflect?
> But in the south,
> By word of mouth,
> It is the opposite, I suspect!

Chapter 6 – High and Low Pressure Systems

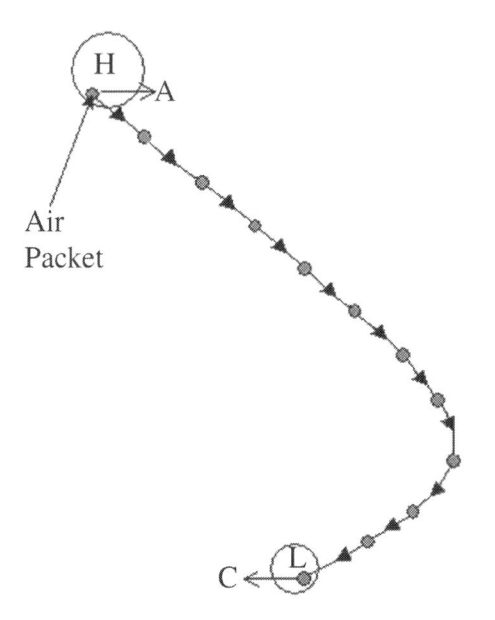

Figure 6.3 The Southern Hemisphere

Here we have the same scenario as in Figure 6.1 only it is upside down – but the air packet as it moves towards the edge of the high is still being deflected to the east and again eventually will over-shoot the low and start to move westward. So, when the air packet leaves the high, heading south it is moving anticlockwise (component A, above), and, when it enters the low, it is moving clockwise (component C, above).

Don't forget, in the southern hemisphere, if the air packet were to exit the high and move north, it would still be deflected to the west, and hence veer anticlockwise (see Figure 6.4).

Understanding the Jet Stream – Clash of the Titans

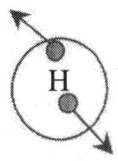

Figure 6.4

6.5 The Measurement of Atmospheric Pressure

Air pressure is measured by instruments called barometers (from the Greek word for weight; they measure the weight of the air). There are two main types: the mercury column barometer, and the aneroid barometer. An Italian scientist, Evangelista Torricelli, designed the first mercury column barometer in 1608. It measures the height of mercury that the air pressure can support in a vacuum. It was found that the average pressure, at sea level, was 29.92 inches of mercury and this became known as "standard atmospheric pressure". Meteorologists use a unit called a *millibar* (mb), where standard atmospheric pressure is equivalent to 1013.25mb.

We must remember that pressure is a force, and, in order to standardise using the *SI* the international system of units (no different than trying force the UK to use kilograms, etc), it has been decided that air pressure should be measured in the SI unit of force – the pascal. Don't be silly it's never that easy! For scaling reasons, they use the prefix hecto, and, like a kilogram, which is a 1000

Chapter 6 – High and Low Pressure Systems

grams and the hectogram is 100 grams, the unit is a hectopascal (hPa), which is 100 pascals, Now, wait for it: 1013.25mb = 1013.25hPa, that's strange they are equal – oh well! Now you know. On the weather charts, which we will look at later, the air pressure is quoted in millibars, and I think it will be quite some time before the SI unit is adopted.

As mentioned earlier, atmospheric pressure is the weight of the air molecules above you, so, as you ascend, obviously the number of air molecules above you is less than at sea level and therefore the air pressure is less. As an example, at the summit of Mount Everest, 8,848 metres above sea level, the average air pressure is 337millibars, about a third of that at sea level. That is why it is usual for mountaineers to carry their own oxygen supply at the summit. This is also the reason why modern airliners have to have pressurised cabins, as the cruising altitude is around 10,000 metres and the pressure is less than a third of that at sea level.

What does all this mean? Well, when air pressure is represented on weather charts, it would be meaningless to show actual pressure. For instance, if a large high pressure covered the UK, and the actual pressure was constant, say 1030mb, then we would want to know this. However, if you were read a barometer in, say, Blackpool (at sea level), you would get a reading of 1030mb, but, a barometer at the top of Kinder Scout (636metres above sea level), in the Peak District, you would get a reading of about 966mb! Not really representative of the actual situation! But the clever weathermen have a trick up their sleeves, they standardise the pressures to what

Understanding the Jet Stream – Clash of the Titans

is known as the Mean Sea Level (MSL) pressure. They use equations into which they enter the barometer readings plus the height above sea level and, bingo! Out pops the pressure as if the reading had been taken at sea level. A rough guide, the one I used to calculate the Kinder Scout pressure (in reverse, I knew MSL pressure, so I estimated what the barometer reading would be), is that for heights up to 1000 metres the pressure drops about 1 millibar for every 10 metres rise. Kinder Scout is 636 metres above sea level and the MSL pressure is 1030mb – so we have:

1030-636/10 = 966mb.

Anyway, I think that's enough, other than to say that in Chapter 9 I will delve further into the mysteries of standard weather charts.

Chapter 7 - What are Weather Fronts?

Weather fronts, or just fronts, is a term conceived by a Norwegian research scientist by the name of Vilhelm Bjerknes, who formulated the concept of fronts by correlating local inclement weather with the boundary where two different air masses met (Sounds familiar? It's a little like the formation of the Jet Stream). As it was just after the First World War, he could see the similarity of the weather systems to two opposing armies facing each other on the battlefield, (militarily known as the front line); hence, he coined the phrase "weather front" to describe this weather conflict. He was a brilliant scientist, who developed weather modelling, and he is considered to be the father of modern forecasting using mathematical models, not bad, considering it was over ninety years ago!

Earlier I said a front is the boundary where two different air masses meet, usually at mid-latitudes, so this implies that the two air masses have different temperatures. Warm air is less dense than colder air, and, therefore, the warmer air will rise above the cooler air – the Montgolfier brothers, in a hot air balloon, used this principle in man's first flight, in 1783. It is no different at a weather front; the warmer air will always rise above the cold. Allowing the colder air to creep underneath it.

I shall briefly describe the associated weather of the three types of fronts, but, for further information and superb diagrams (mine below are OKish), visit the Met Office Website.

Understanding the Jet Stream – Clash of the Titans

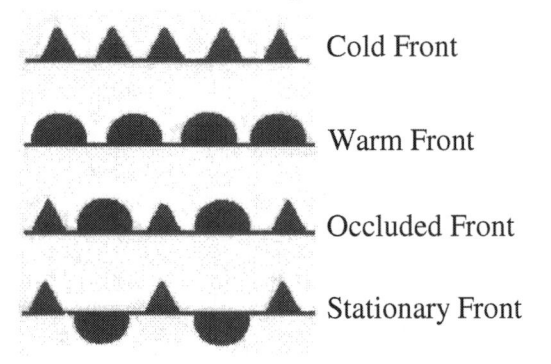

▲▲▲▲ Cold Front

⬤⬤⬤ Warm Front

▲⬤▲⬤ Occluded Front

▲⬤▲⬤ Stationary Front

You may be familiar with some of the above symbols, especially the cold and warm fronts. On weather maps, the triangles are usually in blue, to indicate cold. I remember this because they remind me of icicles, and the hemispheres are in red to show warmth. What they also tell us is where the warm or cold air is, "it's behind you!" Well, not quite; it's on the other side of the symbols, so the symbols are pointing in the direction of the movement. For instance, the cold front above tells us that the cold air mass is below the line and moving towards the top of the page. But what on earth do the occluded and stationary fronts mean? Read on.

In the next sections I mention one or two different cloud types and, in Chapter 12, I have given a brief overview of these and the other major cloud types.

Chapter 7 – What are Weather Fronts?

7.1 The Cold Front

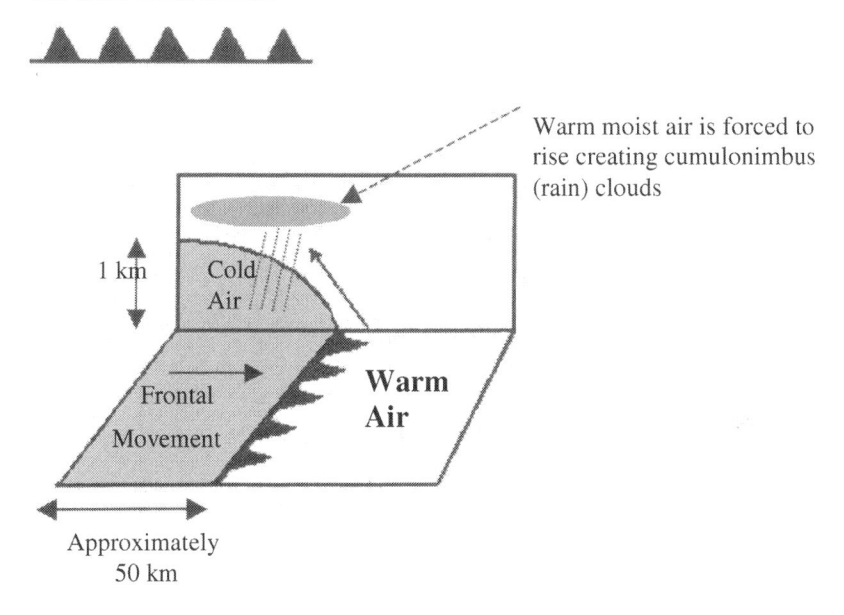

Figure 7.1

Here the cold air is moving from left to right, and, as it moves forward, it has encountered a warmer air mass. Because the warm air is less dense than the cold, the warm air is forced up over the top of the advancing cold air mass. As we are all know the air temperature decreases with altitude, approximately 10°C for every 1000 metres, so, at the top of Ben Nevis (1,344 metres) we would expect the temperature to be just over 13°C cooler that at its base.

So, when warm, moist air rises, it is cooled, and, as the water vapour cools, its molecules slow down. They now do not have enough energy to remain suspended as a gas, so they change back into water droplets- this is called condensation.

Well that's the theory, but water is a strange beasty, so I'll just mention, in passing, that water, in its three states in the atmosphere, before they can change to a less active state (vapour to water and water to ice) requires a catalyst – these are called condensation nuclei and freezing nuclei. These nuclei are just dust, ash, salt, or any other pollutants. For instance, water particles can remain suspended at −40°C (yes, minus 40!) and still not change to ice, unless there is a freezing nucleus to trigger the process. There is a very strange cloud formation called a "Fallstreak Hole" which develops due to super-cooled water vapour. I had a web link for you but, unfortunately, it is no longer available – you can always try to find some photographs on the net.

In any case, after condensing out, the water molecules coalesce and form water droplets and eventually rain or snow.

This is not really a meteorology handbook, but it is worth mentioning, for the devil of it, that things are even more complicated, as you would expect. When there is a change of molecular state - i.e. solid to liquid, or liquid to gas, then energy is required to initiate the change of state, and when it is reversed, gas to liquid, liquid to solid, then energy is released. The energy required/released is heat. What has this got to with the formation of rain clouds? Above, I stated that the temperature drops by about 10°C per 1000 metres – but, when the air rises, some of the water vapour will condense out to water molecules; this is a change of state from gas to liquid so heat will be generated. Perhaps you can now see the complication (but it's not too bad) - all that happens is

Chapter 7 – What are Weather Fronts?

that, instead of 10°C per 1000 metres it only drops by about 6°C per 1000 metres. For further reading check out *stable* and *unstable* air.

Anyway, enough of that. Back to our cold front. If the upper wind speeds are strong enough, then high-altitude cirrus clouds may form ahead of the front (up to 100 kilometres ahead). Otherwise, the first thing you will notice is a fall in temperature followed immediately by rain-bearing clouds (cumulonimbus). It will be all over and done with in less than an hour.

7.2 The Warm Front

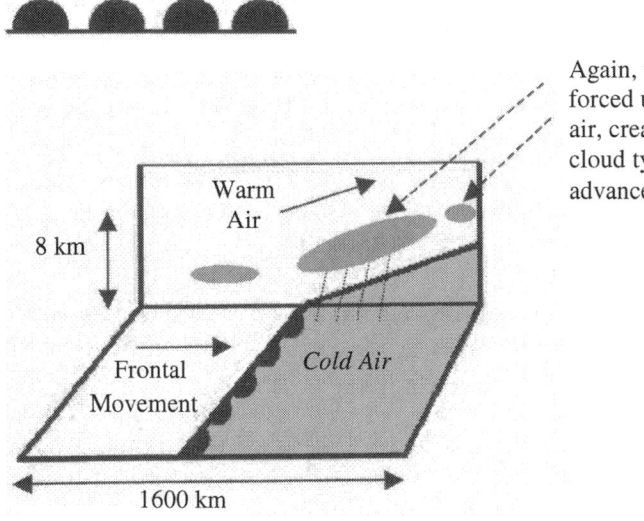

Again, warm, moist air is forced up over the cold air, creating a variety of cloud types well in advance of the front

Figure 7.2

Firstly, notice the scale with the cold front: all the activity is within 120 kilometres horizontally and 1 kilometres vertically. But the warm front is far more extensive: 1600kilometres horizontally and up to 8kilometres high.

If you compare a warm front with a cold front, then you would have to say that the cold front is like a bulldozer, it just pushes everything in front out of the way, whereas the warm front is a gentle giant, just ambling along and gradually rising up and over the cold air ahead of it.

Obviously, the rising warm air will still create clouds, but they will be of a different nature than the clouds formed by the cold front. These are "gentle clouds", mainly stratus and cirrus. There is a

Chapter 7 – What are Weather Fronts?

possibility of nimbostratus forming just ahead of the front, which could bring rain, but, all in all, they are nice clouds.

Most of us have probably witnessed the approach of a warm front without actually knowing it. One of the telltale signs is the presence of high-altitude cirrus clouds; these form well ahead of the front (over a thousand kilometres). As the front approaches, there is a possibility of rain, then a temperature rise as the front passes; this is followed by "cotton wool" stratocumulus clouds. The whole process can take two or three days!

Understanding the Jet Stream – Clash of the Titans

7.3 The Occluded Front (Occlusion)

Before we go any further, we can infer from the previous two sections on cold and warm fronts do not move at the same speed. A cold front moves at about forty kilometres per hour and the warm at twenty kilometres per hour. The frontal ground speed is directly proportional to the density, as cold air is denser than warm air, hence a cold front moves faster than a warm front. When they meet, the denser cold air tends to ride roughshod and just pushes the warm out of the way, up and over.

Have a look at Figure 7.3; here we have a glass tank split into two sections by a sluice gate, which is closed. The left-hand side is filled with water and the right-hand side with treacle.

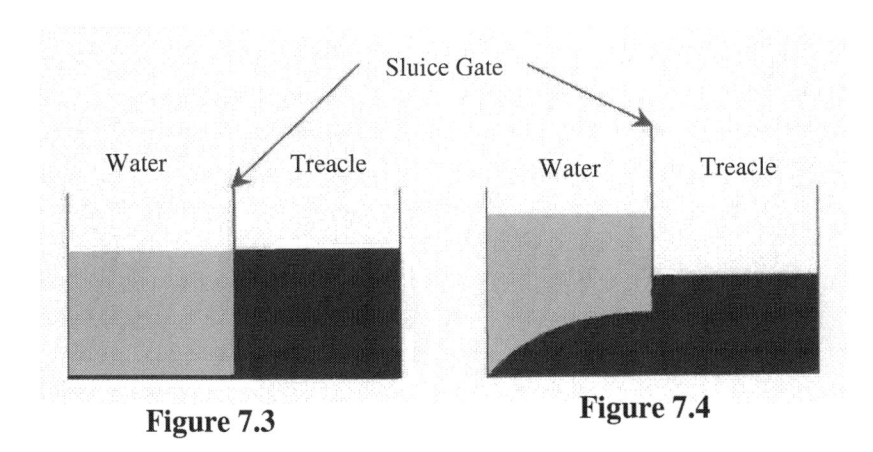

Figure 7.3

Figure 7.4

In Figure 7.4, we have raised the sluice gate and waited a couple of minutes. You can see what has happened: the denser treacle has barged its way into the left-hand side and pushed the water up and

Chapter 7 – What are Weather Fronts?

over it. So, at ground level, the cold front will eventually catch up and push the warmer air out of the way.

If we repeated the experiment, by replacing the treacle with paraffin, which is less dense than water, the reverse would happen: the water would encroach into the right-hand side and push the paraffin out of the way.

The scenario for an occlusion requires that there are three different air masses involved, all at different temperatures. There are two types of occlusions: a warm and a cold occlusion; the difference, I hope, will become clear later.

1 - Formation of a Cold Occlusion
Stage 1

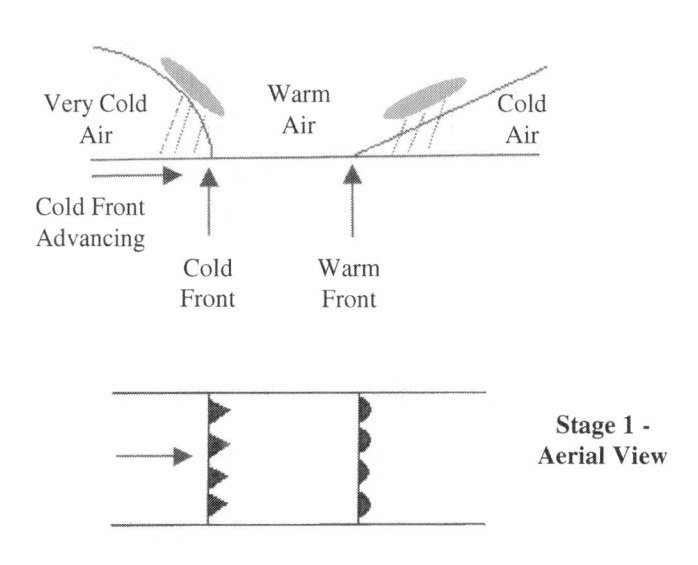

Figure 7.5

Understanding the Jet Stream – Clash of the Titans

What we have here is a very cold front moving left to right and warm air in front of it. Notice that to the right of the warm air is cold air (as opposed to very cold air on the left), so this forms a warm front. Individually, these fronts obviously have the same characteristics as described earlier (7.1 and 7.2).

As the cold front progresses, it eventually catches up with the warm front, as shown in stage 2.

Stage 2

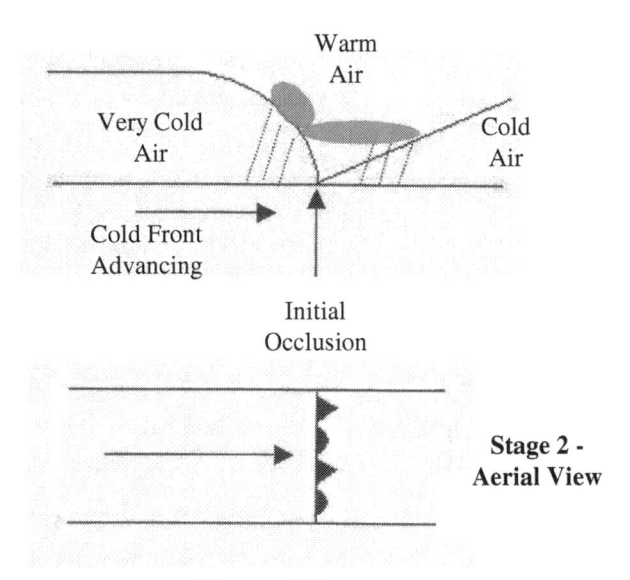

Figure 7.6

The two fronts merge, producing rain both in front and behind the occlusion. Because the very cold air is denser than both the warm and the cold air, it has already lifted the warm air, and, as the very cold front advances, it will start to undercut the cold air and lift it.

Chapter 7 – What are Weather Fronts?

Stage 3

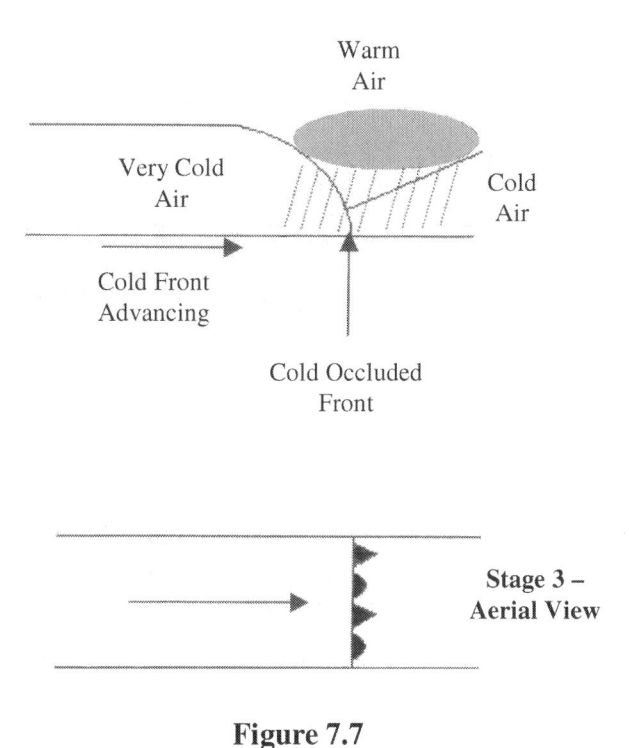

```
                    Warm
                    Air
                   ╱─────╲
   Very Cold     ╱ ▓▓▓▓▓ ╲    Cold
     Air       ╱▓▓▓▓▓▓▓▓▓╲    Air
─────────────╱──╲▓▓▓▓▓▓▓╱──────────
       →         ╲▓▓▓▓▓╱
   Cold Front      ╲▓╱
   Advancing        ▲
                 Cold Occluded
                    Front
```

```
    ────────→  ▶
               ▶     Stage 3 –
               ▶     Aerial View
    ──────────────────
```

Figure 7.7

Here the occlusion is in full swing, producing heavy rain in front and behind the occlusion.

The name of the occlusion is derived from the relative temperatures of the two air masses on either side of the warmer air. If the advancing air is colder than the air ahead of the warm front, then it is a cold occlusion Stage 1, above, shows very cold air versus cold air, so this is a cold occlusion.

47

2 - Formation of a Warm Occlusion
Stage 1

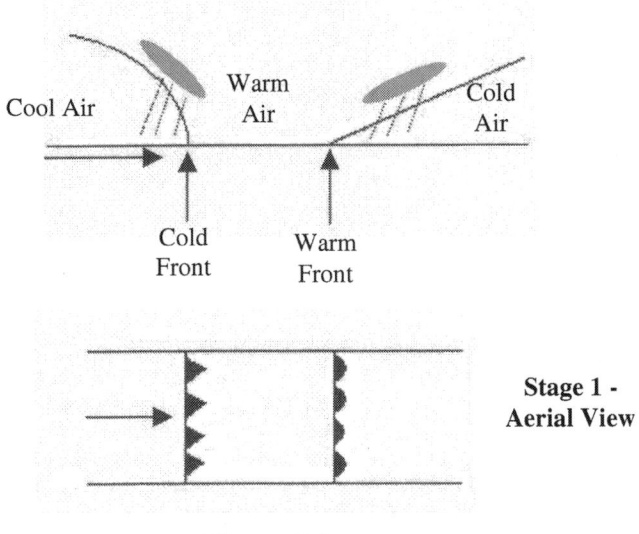

Figure 7.8

This is exactly the same scenario as the cold occlusion, stage 1, except for the temperature of the air behind and in front of the warm air.

It now progresses identically, up to the point of initial occlusion, as for the cold occlusion. Then, instead of the moving front undercutting the cold air, it rises above, as cool air is not as dense as the cold air. So, we now have the following situation.

Chapter 7 – What are Weather Fronts?

Stage 3

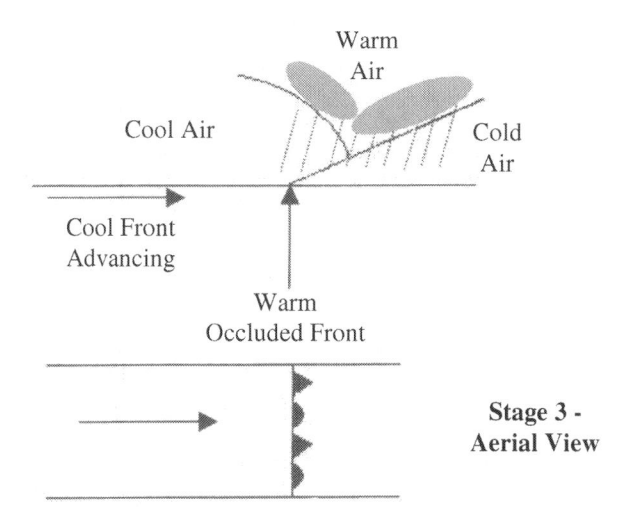

Figure 7.9

That is what occlusions are all about – they generally bring unsettled conditions, with the high risk of rain; heavy rain is associated with the passage of a cold occlusion. Now you are thinking, "So, what has it got to do with the Jet Stream?" I shall just say here that in the next section on frontal lows you will read how these develop and what sort of weather they bring. Here's your clue: a frontal low consists of the passage of a warm front, followed by a cold front, and then, possibly, an occlusion – this spells R-A-I-N in capital letters.

Finally, you may ask, "Can you tell, from a weather chart, what type of occlusion it is?" Unfortunately, the answer is no – this is because

Understanding the Jet Stream – Clash of the Titans
you cannot determine the relative temperatures of the air masses involved.

Chapter 7 – What are Weather Fronts?

7.4 The Stationary Front

So, what have we got here? Remember that, the symbols on a front "point" in the direction of movement, so the symbol above shows that there is a cold front moving up the page and a warm front moving down. Well, yes and no, if this were true: it would really just be a cold front (the cold front being the bully it always is). What it actually represents is that there is warm air above and cold air below and that they are moving parallel to each other but in opposite directions. Phew, that was a mouthful! The name *stationary* is confusing; the two fronts are not actually stationary but moving parallel to each other. What is meant is that they are stationary in their frontal movement (in the direction you would expect then to be moving).

Is that all? No, unfortunately, they are the nursery for the formation of frontal lows (see next the section), and they can stretch for miles. Anyway, much more of this in the next section

7.5 Frontal Lows (Depressions) - Formation

Frontal lows are the bane of our weather, they are synonymous with rain and wind and can sometimes be very destructive (as the UK's Great Storm of October 15/16, 1987). Even worse, they often have an entourage, so you get rid of one and the next one is waiting in the wings.

So, what are these beasts? In Chapter 6, I mentioned the synonyms for low pressure (cyclone or depression). A Frontal Low is a depression formed along a front – easy, eh?

Well, again, no, unfortunately. The formation of a frontal depression is very complex, there are many theories and models to explain this phenomenon, and they are all beyond the realm of understanding for us lay people (you're perfectly welcome to try). But I will have a go and give you my interpretation of (you'll like this!) *cyclogenesis*.

There are two types of cyclones: firstly, the tropical cyclones that form in and around the *tropic*; these are *convective*. Then there are those that develop in the *mid-latitudes*. (Why are these not called *extratropical*, i.e., "outside the tropics"? It is to avoid confusion. An extratropical cyclone is one that has formed in the tropics and has moved out of the tropics into the mid-latitudes. In the UK, we occasionally get the remnants of a deep tropical cyclone, a *hurricane*, which developed around the Gulf of Mexico.)

Mid-latitude cyclones can only form on fronts and so are called frontal – from here on, we are only interested in frontal, mid-latitude depressions.

Chapter 7 – What are Weather Fronts?

Three conditions must be present for the creation of a frontal low; there must be a collision between a cold front and warm front, the fronts must be moving in opposite directions (the definition of a stationary front), and there must be a fast (in excess of 70miles per hour), high-altitude wind. Hang on a minute, high-altitude, fast wind, that sounds familiar! Isn't it the Jet Stream? Sounds like it to me. Aha! Frontal lows can only form beneath the Jet Stream.

But wouldn't frontal lows be regular features of our weather? Again, there are three reasons why not: firstly the Jet Stream is not always in the proximity of the UK, and the temperature gradient (difference in temperature of the air masses on either side of the front) and the speed of the Jet Stream must both be above critical values. However, they are frequent visitors to our shores.

I am not even going to try to draw the development of a frontal low but, instead, I will show you the evolution, using actual MSL pressure charts. In Chapter 9, I will explain these in more detail, so please don't worry about everything else on the chart; just concentrate on the frontal low's progression. The charts I have included are over a period of three days - starting at 12:00 hours *UTC* on 02 November 2008 - in twelve-hour intervals.

Understanding the Jet Stream – Clash of the Titans

02 November 2008, 12:00 UTC, Stage 1

Stationary Fronts   **Figure 7.10**

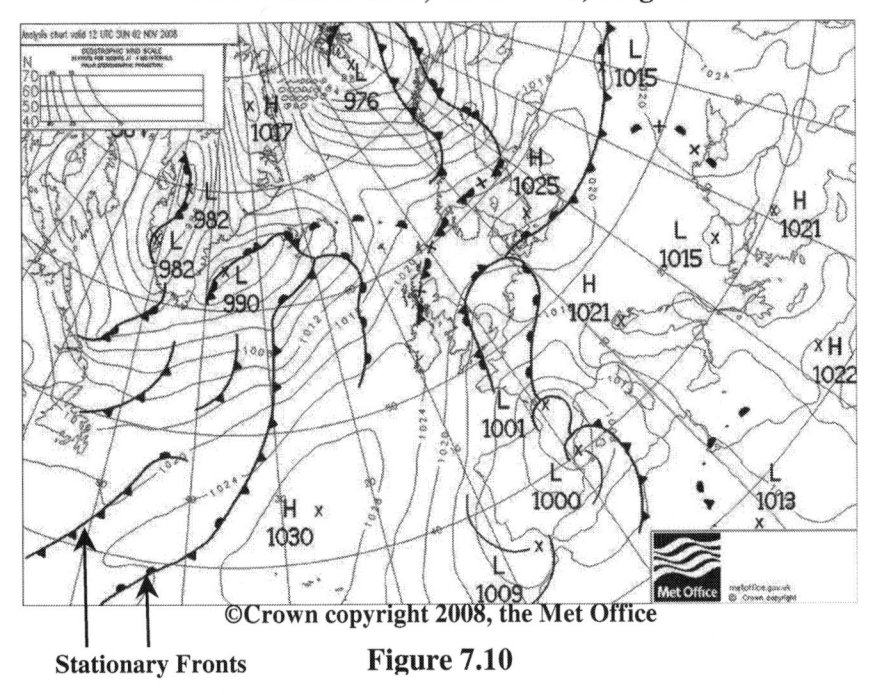

Figure 7.10 shows two stationary fronts, on which a frontal low may or may not form.

Chapter 7 – What are Weather Fronts?

03 November 2008, 00:00 UTC, Stage 2

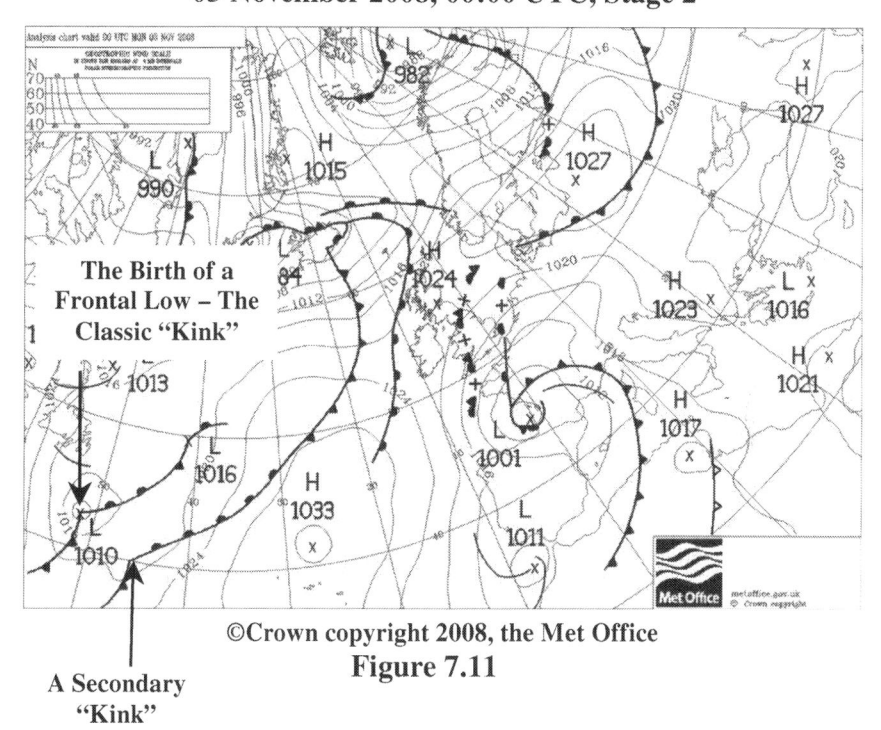

©Crown copyright 2008, the Met Office
Figure 7.11

Lo and behold, as if by magic, the telltale initial "kink" has appeared (how did I know that was going to happen?). Notice that a low has already developed: an L marks a low pressure, and the X marks the centre of the low. (The chart also shows us that the pressure at the centre is 1010millbars much more on this in Chapter 9.) These events raise two questions:

1) Bearing in mind that a stationary front can be many hundreds of miles long, why did the frontal low decide to develop at that particular point? Was it purely random, or was there a reason? I could not find a satisfactory answer, so I delved into the archives for a clue. The first sign of the frontal depression often happens in the region of a sharp change of

55

direction of the Jet Stream, but I think it is wise to assume that they are random, in the sense for prediction. They form at a point where the criteria (temperature gradient and jet speed) exceed the critical values.

2) How did the central low pressure area form so quickly? Once critical values are exceeded, conditions become perfect for the development of a low. Remember that a powerful Jet Stream is above the action, and a simplified version of this complex procedure is: the Jet Stream is powerful enough to drag air out of the air column below it, pull it upwards, and spit it out (divergence). This creates a momentary vacuum, and, as described in 5.3.3, this will start the process of convergence at lower levels, and hence the formation of a depression. The "kink" is the first sign, because, as soon as surface convergence begins, the big bully (cold, dense air) barges its way towards the point of convergence (see treacle/water, section 7.3). Keep in mind that the frontal lines and isobars represent the situation on the surface.

Have another look at the first chart; the low was not present, just the kink. So, the divergence/convergence process (described above) was in full swing during the previous twelve hours. Notice, also, that the secondary kink does not have a well-defined low pressure centre.
Let's see what happens next:
The primary kink, above, is located at approximately 42° north and 51° west and the secondary one is at 40° north and 45° west. Figure

Chapter 7 – What are Weather Fronts?

7.12, below, shows the position of the Jet Stream on the same date and time (I can hear you screaming, "It's about time we saw something of the Jet Stream!") Don't get too carried away, by the end of the book you will be fed up with seeing them!

A word of warning: be careful when comparing the two charts (MSL and Jet Stream), as they are on different scales (I have tried rescaling them, but one or the other ends up blurred, due to resolution problems). The arrows point to the approximate locations of the kinks.

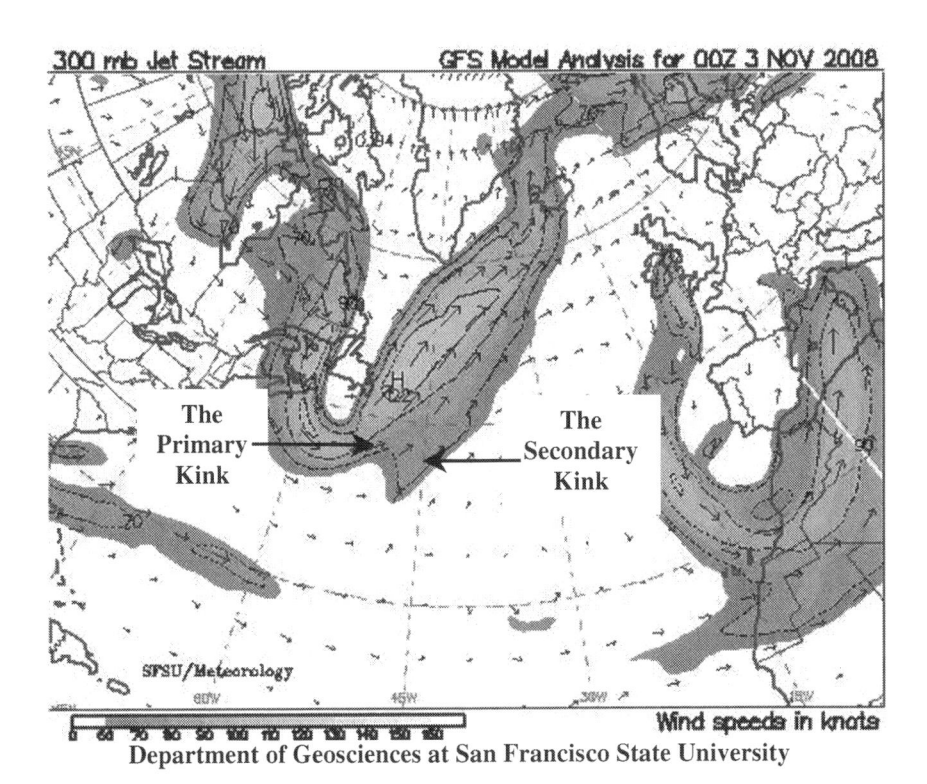

Department of Geosciences at San Francisco State University

Figure 7.12

Understanding the Jet Stream – Clash of the Titans
03 November 2008, 12:00 UTC, Stage 3

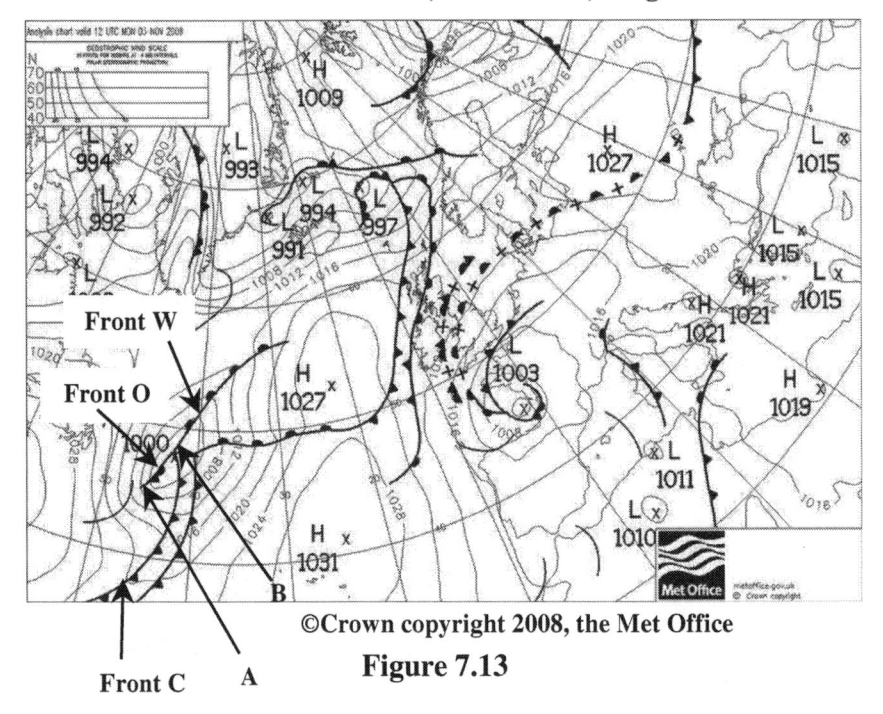

©Crown copyright 2008, the Met Office
Figure 7.13

Figure 7.13, is the next stage - notice that the low pressure centre is now down to 1000mb and covers quite a large area. We are not only interested in the low but also its position. I think you will be surprised at the speed at which these lows move – see Table 7.1

You can see front C is a cold front, W a warm front and O an occluded front (an occlusion).

The occlusion has formed because the cold front C is swinging round anticlockwise and also pushing left to right. As discussed in 7.3, the cold front is moving faster then the warm front, and, between points A and B, front C has barged underneath the warm front and forced it up and over, forming the occlusion.

Chapter 7 – What are Weather Fronts?

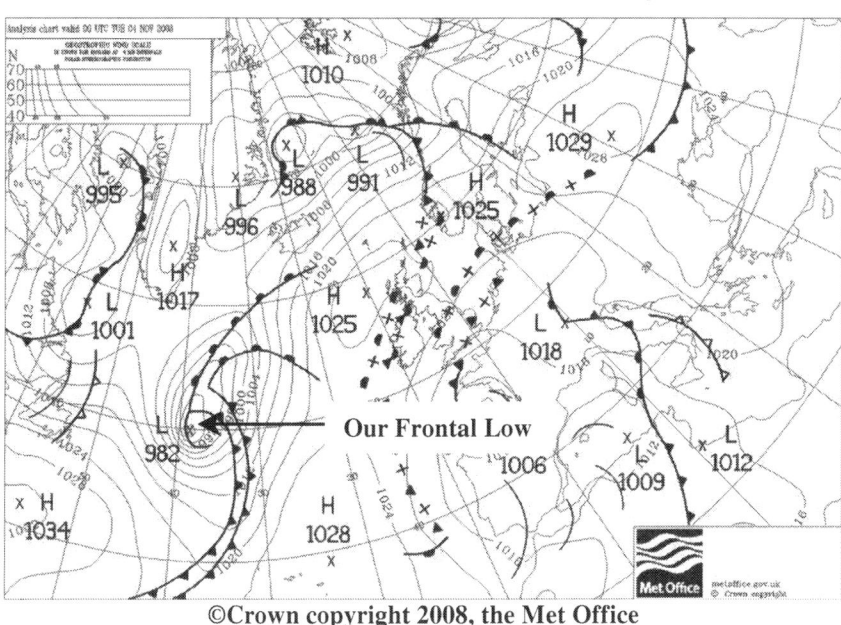

04 November, 2008 00:00 UTC, Stage 4

©Crown copyright 2008, the Met Office

Figure 7.14

Here you will notice that the central pressure has now dropped to 982mb. It also appears that the occlusion has disappeared; I am pretty certain that it is still there but just has not been shown.

I have not mentioned anything yet about the *isobars* around the low, but, if you look back, you will see that they are increasing in numbers around the low as the low deepens. More on this Chapter 9.4, but I will just say that the closer they are packed, the stronger the winds, and as long as the low is deepening, then the wind speeds will continue to increase.

Understanding the Jet Stream – Clash of the Titans
04 November 2008, 12:00 UTC, Stage 5

©Crown copyright 2008, the Met Office

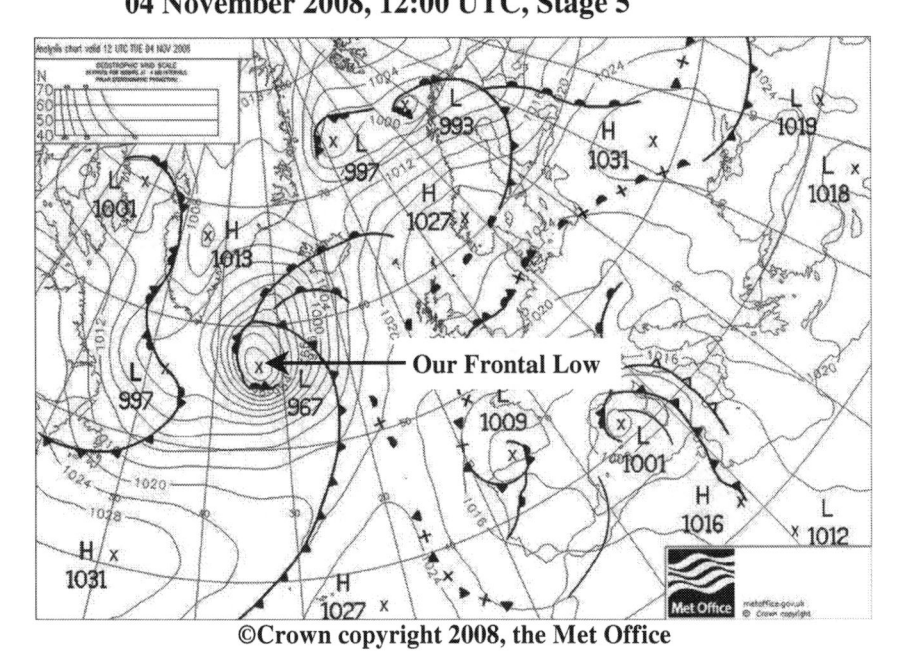

Figure 7.15

Again the pressure has dropped to 967mb.

Chapter 7 – What are Weather Fronts?

05 November 2008, 00:00 UTC, Stage 6

©Crown copyright 2008, the Met Office

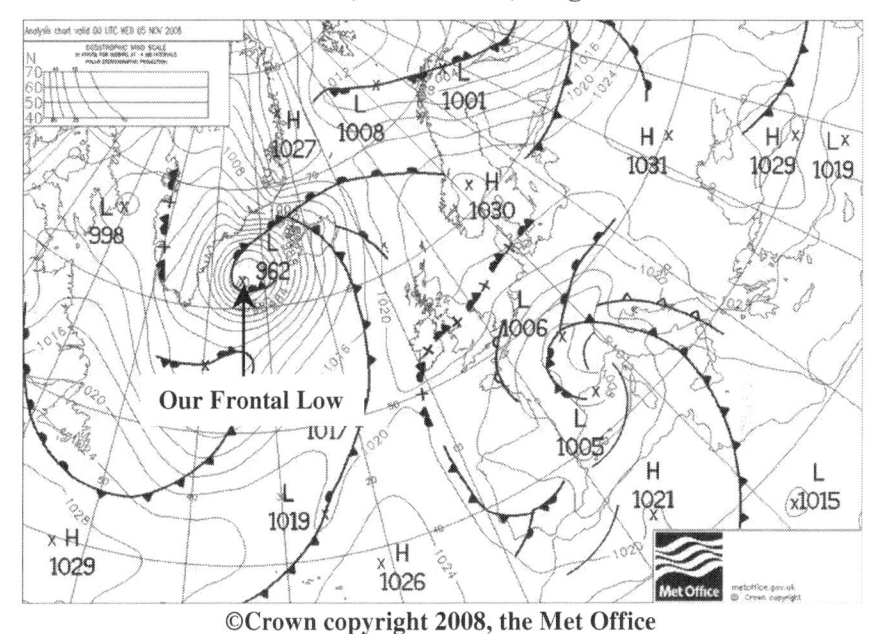

Figure 7.16

Pressure is now down to 962mb with an incredible set of packed isobars. This spells trouble; luckily, the low is not near any major populated areas.

Understanding the Jet Stream – Clash of the Titans

05 November 2008, 12:00 UTC, Stage 7

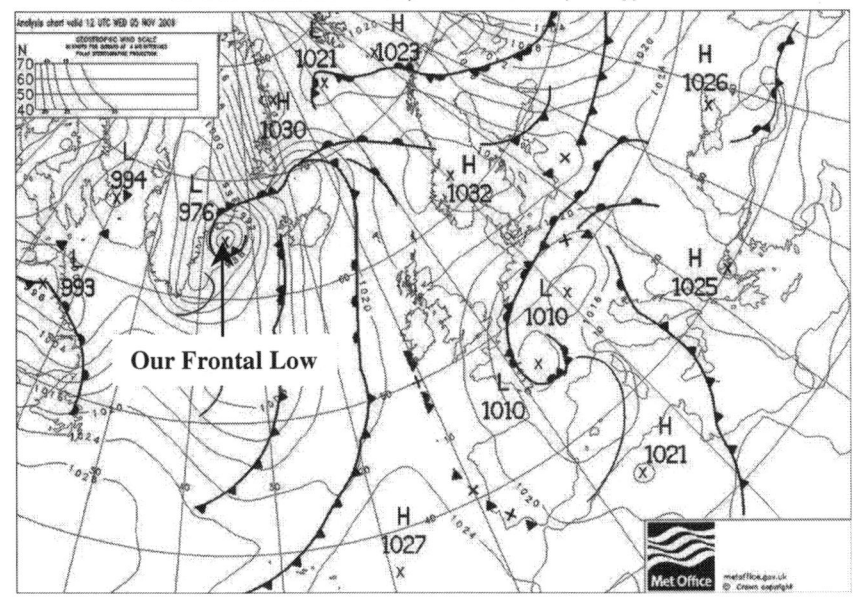

©Crown copyright 2008, the Met Office

Figure 7.17

The first good sign, is that the pressure has started to rise, it's at 976millibars. Notice that it is no longer an organised system; it has fragmented and is losing its structure.

So, from the original kink to the collapse the frontal low took 2.50 days (60 hours). Table 7.1 shows the location of the centre of the low at each stage: longitude in degrees west, latitude in degrees north, and the distance travelled over each 12 hour period, a total of 1969 miles (an average speed of 32.82 mile per hour)!

Chapter 7 – What are Weather Fronts?

Stage No.	2	3	4	5	6	7
Longitude	53	44	39	34	35	35
Latitude	42	48	50	56	63	65
Distance Moved (Miles)	-	608	267	466	488	140

Table 7.1

Incredible amounts of energy are associated with frontal lows (mainly due to powerful winds). Luckily, this depression did not reach the UK; it died out over Greenland, unlike the Great Storm of October 1987, which was caused by a similar frontal low except that its origin was in the Bay of Biscay.

Below are thumbnail extracts of the shape of the fronts from each of the stages above. These are typical of the life cycle of frontal depression.

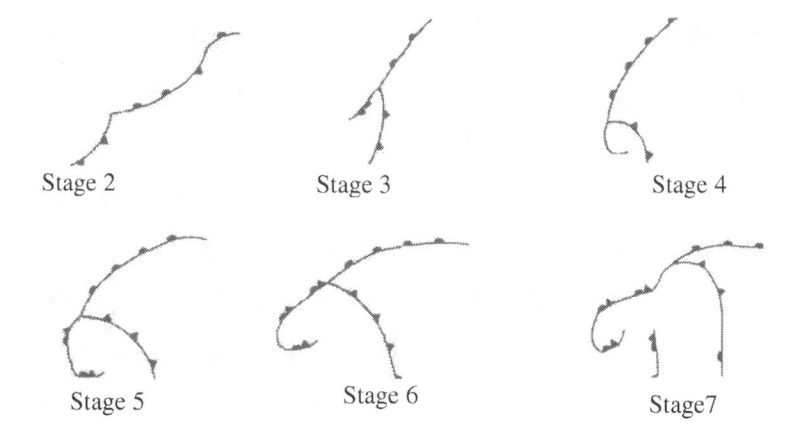

Stage 2　　　　　Stage 3　　　　　Stage 4

Stage 5　　　　　Stage 6　　　　　Stage 7

63

Understanding the Jet Stream – Clash of the Titans

Just to tidy a few loose ends before we have a closer look at how the Jet Stream affects our weather.

There is a classic case of "the chicken or the egg" scenario. We know that our Jet Stream is formed at the boundary between the north polar air mass and the north subtropic air mass, but the Jet Stream is sometimes quoted as having presence in its own right (e.g. the Jet Stream is bringing in cold air from the north). But you must always remember: if it weren't for the interaction of the two air masses ("Clash of the Titans") it would not exist. So, although the Jet Stream appears to steer the lows, the responsibility lies with movement of the air masses. But, importantly, the Jet Stream does show us where the boundary is between the air masses.

If you take a look at the development and track of frontal lows, you will frequently see that they emerge from under the Jet Stream (to the cold side of the front) but without losing (even gaining) strength, so, what is going on? Doesn't a frontal low need the Jet Stream to keep it going?

The Jet Stream is only required to initiate (kick-start) the formation of the low; the energy required to sustain and perpetuate the low is obtained by convection from the moist, warm air of the Atlantic Ocean.

Unfortunately this opens a "can of worms"; as the processes involved would require doubling the size of this book, but I can give you some key words if you would like to delve further into it:

- ➤ Stable and Unstable Air
- ➤ Convective Cloud Formation

Chapter 7 – What are Weather Fronts?

- Ocean and Air Temperatures including the Gulf Stream
- Hurricane Life Cycle

Chapter 8 - Common Jet Stream Profiles Affecting the UK

As mentioned in Chapter 3, the Jet Stream does not have any constant physical properties other than its general direction, i.e. west to east.

Around the UK, the Jet Stream can be present in many patterns and shapes (profiles), which, depending on the time of year, can affect our weather in completely different ways.

For instance, if the Jet Stream is absent (underneath or above us), this means that we are totally at the mercy of the prevailing high and low pressure systems; these can originate from the Atlantic, near continent or the north (Polar). Take, for instance, near continent pressure systems (usually these are high pressure areas); in the absence of the Jet Stream there is nothing to stop them infiltrating our "air space". So, in the summer, we can expect very warm, dry weather from the continent (summer 2006), but in winter we would expect very cold, dry conditions, as the air originates from the cold middle Europe and the Siberian ice fields (January 2009).

The names I have given to these profiles either describe the general direction from where the Jet Stream has come (e.g. the Atlantic or Northerly) or a recognisable shape (the Arch or the Horseshoe).

Firstly, let me give you a crash course on interpreting the Jet Stream charts I have included in the book. There are many sites that show Jet Stream charts but I have found the San Francisco State University site is extremely good. The reason I rate them top is that

Chapter 8 – Common Jet Stream Profiles affecting the UK

their five-day forecast is quite accurate, they have an extensive archive (dating back to March 2006) and they are very helpful (via email). Here is their Web address:

www.squall.sfsu.edu/crws/jetstream.html

Let's have a look at one of their charts:

8.1 Your Starter for 10

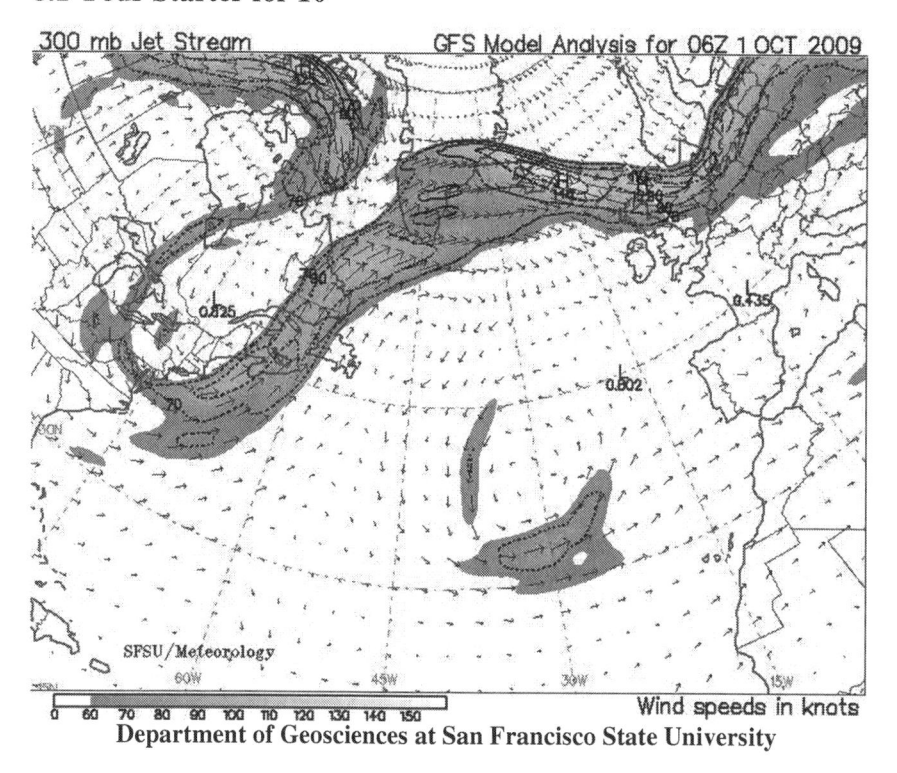

Figure 8.1 - Sample Jet Stream Chart

I have included this chart for no other reason than that it is the current chart at the time of writing. I have also included today's MSL pressure chart, for comparison, later, in Chapter 9.

Understanding the Jet Stream – Clash of the Titans

So, let's have a look at it, starting with the top line "300mb Jet Stream". In section 6.5 you read that air pressure decreases with height and that at sea level the average pressure is 1013mb, so what this is saying is that the chart is showing the wind speeds where the air pressure is 300mb (less than a third of that at sea level). This equates to a height of about 10,000 meters.

GFS – Global Forecasting System - an international computerised meteorological forecasting system.

The shaded areas highlight where the wind speeds are above 60 knots. The scale at the bottom left-hand side shows that it goes from dark grey, 60 knots, through to white, greater than 150 knots.

Be careful not to get confused; the Jet Stream over Iceland shows several shades of grey through to almost white – this says that the wind speed at the centre is about 140 knots. However, the white area inside the larger shaded area off the northwest coast of Africa is below 60 knots. If this white area were to indicate speeds greater than 140 knots, it would mean that the wind speed had literally jumped from 60 knots to 140 knots over an infinitely small distance, which is impossible. But if there is a gradual change from dark grey, through light grey, to white, then this shows winds greater then 150 knots; otherwise they are less than 60 knots.

You will see numbers dotted around; these are the wind speeds (in knots) in that area. Also the letters H and L appear in places; do not get confused. They do mean High and Low, but *not* pressures; again, they are wind speeds. For example, off the west coast of Spain

Chapter 8 – Common Jet Stream Profiles affecting the UK

you'll see L 0.802; at that point, the wind speed is 0.802 knots (less than a mile an hour).

Notice also all the arrows. (A man got stopped for driving the wrong way down a one-way street; the policeman asked, "Didn't you see the arrows?" – to which the man replied, "No, I didn't even see the Indians!"). Enough of that. The arrows show the direction of the wind (at 10,000 metres), and also indicate the wind strength; the longer the arrow, the higher the speed. I find them more useful outside the shaded areas – so, if you again, have a look at the L 0.802 area (in Figure 8.1), you'll see that the arrows are virtually nonexistent; whilst through the shaded area off Africa they are quite long. Notice that, in the region surrounding this shaded area, the arrows are definitely rotating anticlockwise. What do you think is happening here? Alfred Hitchcock, eat your heart out – I'll tell you in Chapter 9!

With each profile, I will give a brief explanation of what the general properties are of the air mass being brought in (see Chapter 4) and the type of weather you would expect.

8.2 The Atlantic Jet

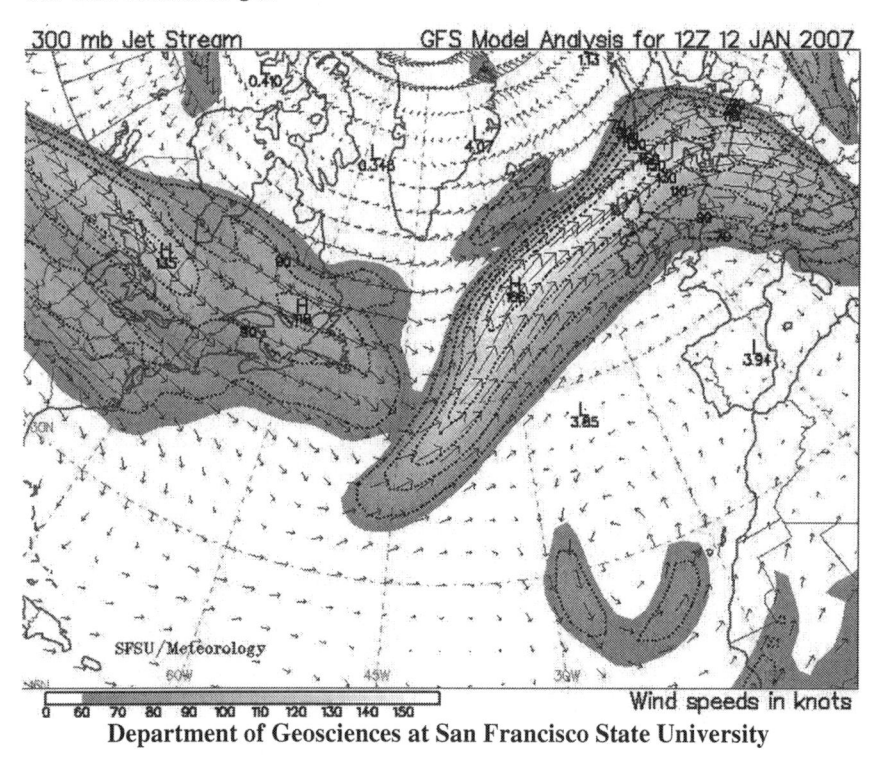

Figure 8.2 – The Atlantic Jet (12 January 2007)

In this example, the air has travelled a long distance over the sea, so it will contain a lot of moisture. Also, the air will be relatively warm (both in summer and winter), as it originated from warm latitudes. It is, however, the perfect cocktail for cloud formation and rain, when it makes landfall. So, in the winter, it will bring us relatively mild air, though rain as well (you can't have it both ways). Also, importantly, it helps to block really cold air from the Arctic or the continent from reaching us. Therefore, in the winter the Atlantic Jet

Chapter 8 – Common Jet Stream Profiles affecting the UK

is a good guy. January 2007 was very mild but there was above average rainfall.

As I have just said, you can't have your cake and eat it, too. In July 2009 Atlantic Jets passed over the UK on eleven (not consecutive) days, this brought more than the average rainfall for July, and it may go down as the wettest July on record (my tomato plants are still trying to recover). But temperatures were average. In summer, we see, the Atlantic Jet is a bad guy. (See Chapter 7.5 - Frontal Lows.)

On some occasions, the Atlantic Jet may generate little or no rain. For instance, the incoming air could have "rained itself out" before reaching the UK. See Chapter 10 on Forecasting Using the JS and Atlantic Charts.

This is a postscript that I felt I had to include; the time of writing is a week after the six-day deluge over Cumbria, in which numerous rainfall statistics were broken. For the whole of the period northwest England and Scotland were under a powerful Atlantic Jet, which was bringing in very moist warm air from the Atlantic Ocean north of the Caribbean Sea, and we know what that means: rain, rain, and more rain – not that we could do anything about it!

The next two profiles, the Northerly and the Southerly, are "relatives" of the Atlantic Jet, but with slightly different genes. All three can have a direct impact on our weather, temperature, and moisture content of the air being brought in. If it is a fast jet (in excess of 90 knots), it can influence the prevailing high and low pressure systems (PHL) and hence the wind direction.

8.3 The Northerly Jet

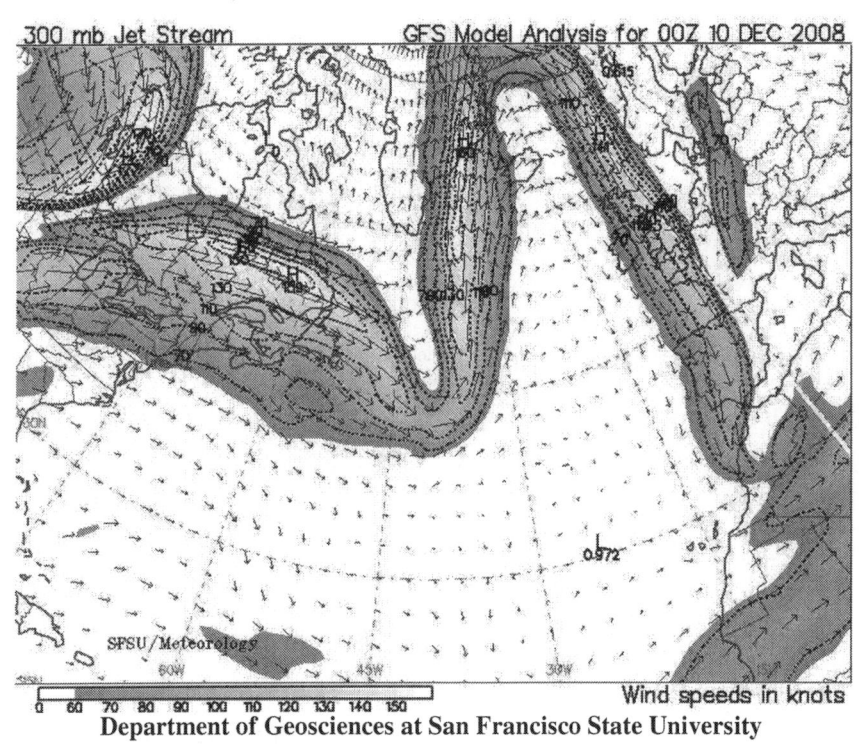

Figure 8.3 - The Northerly Jet (10 December 2008)

Here, in figure 8.3, we have a completely different animal, the obvious distinguishing feature is the source of the air covering the UK. Not only has it originated from colder climes (Canada), but it has swung up and spent time inside the Arctic Circle. If this immediately brings to mind c-c-c-c-cold you would be spot on. Needless to say that the air is cold and dry. December 2008 was the coldest for over a decade and yet rainfall was 50 per cent of the average.

Chapter 8 – Common Jet Stream Profiles affecting the UK

In summer, it is unusual for the Jet Stream to come from the Arctic, but often, high-altitude winds just below the threshold of 60 knots can affect the UK. These always bring unseasonable cold northerly winds accompanied by generally unsettled conditions (12 to 15 June 2008). The Northerly Jet is not a very nice guy.

8.4 The Southerly Jet

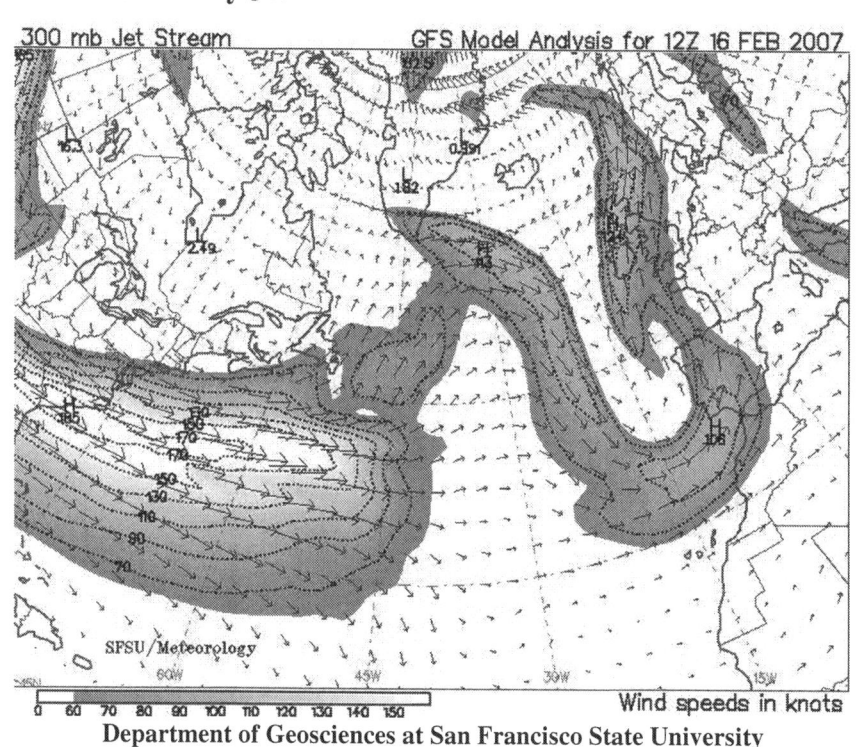

Figure 8.4 – The Southerly Jet (16 February 2007)

In this example, the air has started in a relatively warm zone, passed over cooler waters and then nose-dived to warmer waters and up again, eventually reaching the UK. So, what can we say about the temperature and humidity of the air when it reaches us? Well, this is winter, so it will be relatively warm air and quite moist, because it has spent a lot of time over the oceans.

So, generally, we would expect the temperatures in both summer and winter to be warmer than the average.

In winter, unsettled weather with a good chance of showers, 16 February 2007, was described, by the Met Office as "some heavy rain with temperatures remaining above average". The Southerly Jet then is quite a good guy.

In summer, pleasant to warm with the possibility of localised thunderstorms, an example was the first few day of July 2008 when it was described as "fine and hot with heavy rain and thunderstorms". Good guy, apart from the storms.

The next three profiles differ from the previous three, in that the Jet Stream does not directly influence our weather, temperature or air moisture content, and wind speed and direction are governed by the PHL. But you will see that the position of the Jet Stream and time of year does dramatically affect the weather.

Chapter 8 – Common Jet Stream Profiles affecting the UK

8.5 The Arch Jet

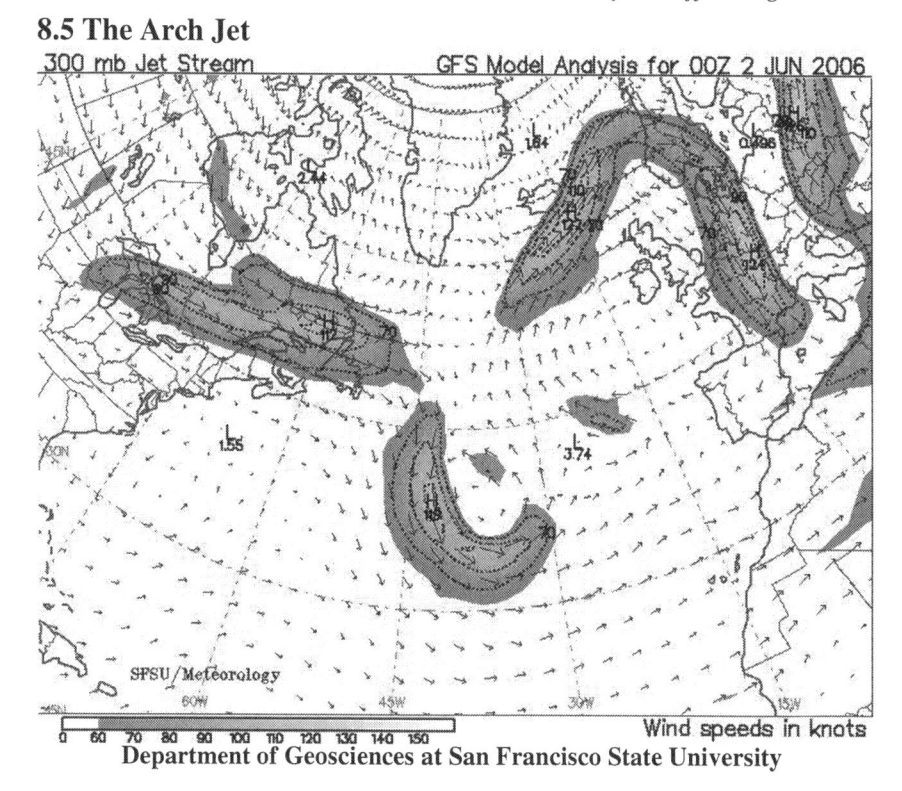

Figure 8.5 - The Arch Jet (02 June 2006)

Since, in figure 8.5, the Jet Stream is "broken," the actual source of the air is not as obvious as that of the Atlantic Jet. But I think it is reasonable to assume that it originated from the near continent after spending some time over the mild Atlantic (the movement round the high pressure system off the north coast of France). Hence the air covering the UK is warm and moist. As usual, there is always a down side. In the summer, when temperatures are warm to hot, the moist air can generate localised convective clouds – thunder, lightning and heavy rain. The first ten days of June 2006 were described by the Met Office as "warm with occasional

Understanding the Jet Stream – Clash of the Titans

thunderstorms". So, in the summer the Arch Jet is an OK type of guy.

In the winter, because the UK is bathed with relatively warm, moist air, we should expect better-than-average temperatures, settled weather, and dryness. Archives show an Arch Jet Stream on the 1, 2, 3 and 4 February 2007, and the Met Office described February 2007 as having a "dry and fine start to the month". So, a good guy.

8.6 The Horseshoe Jet

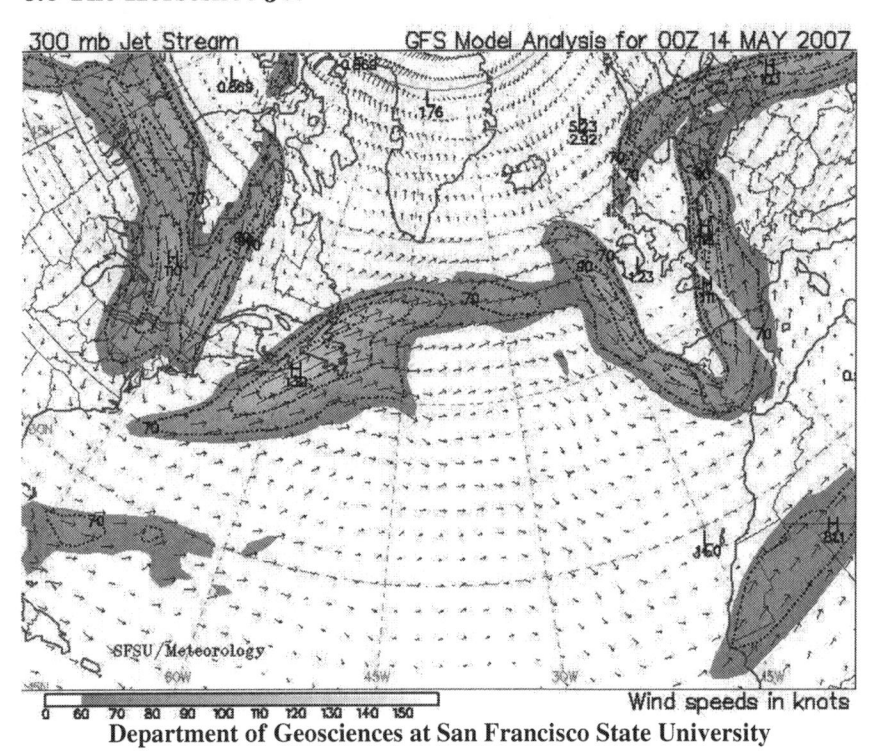

Figure 8.6 - The Horse Shoe Jet (14 May 2007)

Here the source of the air is from the warm western Atlantic, meaning warm and moist air. We would, therefore, expect similar

Chapter 8 – Common Jet Stream Profiles affecting the UK
conditions as the Arch, both in summer and winter. However, because the UK is north of the Jet Stream, this could slightly reduce the temperatures, depending on the local winds, as they could be from the north.

So in summer and winter we would expect the weather to be damp, unsettled and possibly milder than the seasonal average. So, the Horseshoe Jet is an OK type.

8.7 The No Jet

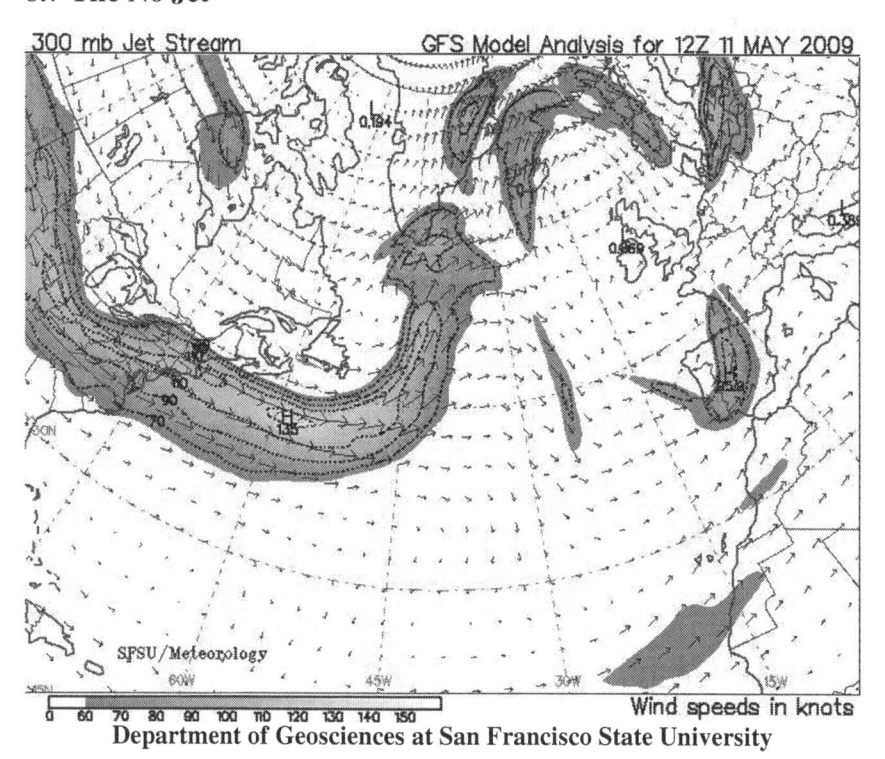

Figure 8.7 – No Jet (11 May 2009)

In the summary below, you will find, under the column heading W, the abbreviation PHL; as mentioned before it means "Prevailing

Understanding the Jet Stream – Clash of the Titans

High and Low Pressure systems". In the No Jet scenario, Figure 8.7, our weather is totally dependent on the prevailing high and low pressure systems. On 11 and 12 May 2009, figure 8.7, a high pressure system developed over the UK (notice the L over Northern Ireland has a wind speed of 0.869 knots). It brought pleasant warm conditions with little rain. The skies were clear, which meant a cold night. Good guy.

The first few days of February 2009 were also affected by No Jet conditions; this meant that a Eurasian high (Table 4.1) could take control, and, as expected, it was very cold. However, when this cold air reached our shores, which are relatively warm and mild, the inevitable happened: heavy snow. The Met Office reported "The first half of the month was very cold with some heavy snowfalls". Not such a good guy (unless you are a kid at heart).

The type of weather expected with No Jet describes what our weather would be like if the Jet Streams did not exist; extreme, predictable, and boring!

Just another point to remember: the UK is small, but it is big enough for different regions to be affected by the Jet Stream in different ways. If a Jet Stream covers the whole of the UK, then the chances are we all get similar weather (bearing in mind the Jet Stream speed). But, for instance, in early September 2009, the Jet Stream was predominantly an Arch, and the arch crossed the north-west of Scotland. South of the border, it was fine and mild, whereas Scotland had heavy rain with strong westerly winds and displayed the characteristics of an Atlantic jet.

Chapter 8 – Common Jet Stream Profiles affecting the UK

8.8 Summary

Below, table 8.1 summarizing the type of conditions expected from the 6 profiles:

Jet Type	Winter		
	T	P	W
Atlantic	Milder	Yes	Westerly
Arch	Milder	Some	PHL
Horseshoe	Cooler	Yes	PHL
Northern	Colder	No	Northerly
Southern	Milder	Yes	Southerly
None	Colder	Some	PHL

Table 8.1a

Jet Type	Summer		
	T	P	W
Atlantic	Cooler	Some	Westerly
Arch	Warmer	Thunder	PHL
Horseshoe	Milder	Yes	PHL
Northern	Cooler	No	Northerly
Southern	Warmer	Some	Southerly
None	Warmer	Some	PHL

Table 8.1b

Key to Tables 8.1:

- **T**   Temperature compared with average
- **P**   *Precipitation*
- **W**   Wind direction e.g., Northerly covers NW through North to NE; Easterly covers NE through East to SE; Southerly SE covers through South to SW; and Westerly covers SW through West to NW. The direction should always be confirmed by the PHL, especially if the jet speed is low.
- **PHL**   Wind direction and speed determined by the Prevailing High/Low Pressure systems. This can result in revised temperature and precipitation forecasts.

Understanding the Jet Stream – Clash of the Titans

You should remember that the Jet Stream indicates the boundary between the air masses and that the first three profiles, Atlantic, Northern and Southern, means the UK is "straddling" these air masses, the North Polar and the North Subtropic; which implies unstable conditions. Whereas, for the Arch, Horseshoe and None, the UK is in either one or the other air masses, which implies more stable conditions.

Here is another one of my buts; the profiles I have chosen are based on the best examples of each type with well-defined shapes and characteristics. For instance, my Northerly Jet ticks all the boxes we would expect; it is blowing almost directly from the north and has actually spent time over the Arctic, and so it will be cold and dry.

However, when you take a look at your first Jet Stream forecast map, it is unlikely that you will be able to say, "Ah, on such-and-such a day we are going to be affected by an Arch Jet Stream". But, what you must do is get a feel for the general shapes and, most importantly, the source region and the route to the UK of the air mass plus any relevant high and low pressure systems. (It is equally unlikely that you will ever see an identical profile to ones included here.)

You will have noticed a difference in the widths of the Jet Stream during the winter and summer months; take a look at Figures 8.2 (January 2007) and 8.7 (May 2007). The winter jet generally exhibits higher speeds and greater widths than the summer jet, due

Chapter 8 – Common Jet Stream Profiles affecting the UK
to the fact that the Jet Stream is more vigorous in the winter than the summer. This is because of temperature differences (temperature gradients) between the air masses. In the winter the polar air is very cold, and the subtropic air is mild to warm, but, in the summer, the polar is cool to cold and the subtropic is air only a little warmer than in winter. To demonstrate this strength, think of a house with a gap under an outside door. In the summer, the difference between inside and outside temperatures is small (a few degrees either way), so there will be little or no draught. But in the winter, the difference in temperatures is considerably higher, perhaps in excess of 20°C; now you will feel a very strong cold breeze under the door (hence the draught excluder).

Just as an aside, when visiting the San Francisco State University site, it is worth having a look at the whole of the northern hemisphere (top left-hand side of the page), it shows where the Jet Stream originated, its progress round the planet over Europe and into the Arctic circle.

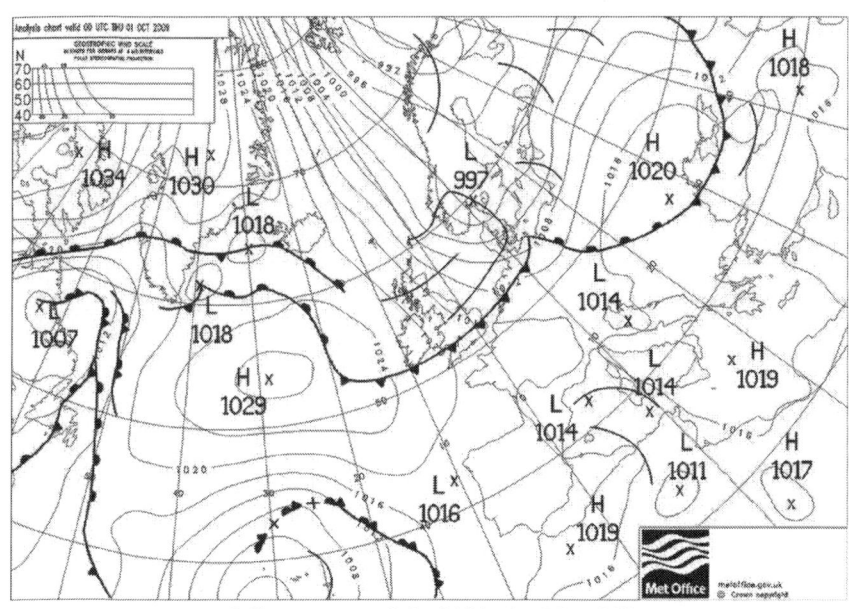

Understanding the Jet Stream – Clash of the Titans

Chapter 9 – An Introduction to the Weather Charts

©Crown copyright 2009, the Met Office

Figure 9.1 –

The MSL Pressure Chart Corresponding to Figure 8.1

The first point I should mention is that the date and time of this chart is 00:00 hours on 01 October 2009, but the corresponding Jet Stream chart (Figure 8.1) is six hours later (06:00 hours), so there will always be a slight discrepancy.

This chart shows the North Atlantic and is often referred to as an Atlantic Chart. You can see quite clearly the outlines of a little bit of Canada, Greenland, Iceland, Great Britain, Europe, and a wee bit of North Africa.

Chapter 9 – An Introduction to Weather Charts

The thick black lines represent fronts, as described in Chapter 7 and you should be able to recognise them.

The only exceptions are the thick lines without any symbols, e.g. west to east over Scotland – these are troughs. These are defined as "an elongated region of relatively low pressure" they are supposed to indicate an increased risk of cloud and rain. Well, I have seen a trough, clearly marked as heading my way, and waited, and waited, for something to happen, and you know what – nothing happened, so I'm not going to say anything more about them.

The remaining thinner lines are either isobars (see 9.4) or lines of longitude and latitude, both in 10° increments. If you are unfamiliar with these, have a look at Chapter 1.1.

In the next few sections, we are going to have a look at some specific points of interest on the chart above and on Figure 8.1, plus other charts of interest.

9.1 Low Pressures

Have you solved the Hitchcock suspense (Chapter 8)? Yes, good - it is an area of low pressure (remember, low pressure systems rotate anticlockwise in the northern hemisphere).

At first glance, you may think that it is the L off Spain's coast, but it isn't - don't forget the scale of the two charts is different. The centre of our low is situated just off the chart at about 31° west and 34° north.

Understanding the Jet Stream – Clash of the Titans

If you check the five surrounding isobars (the outer most one is 1016mb) you should be able to work that the centre of the low is about 997mb.

Take a look at the low, latitude 61° north, longitude 39.5° west. Firstly, you'll see that the pressure is 1018mb but notice the kink; this tells us that it could be the start of a frontal low.

Figure 9.2 is the Atlantic chart two days later (03 October). You will see that it developed into a full-blown frontal depression, which brought very heavy rain and gale-force winds to Scotland and the border areas and very strong winds over the rest of the UK.

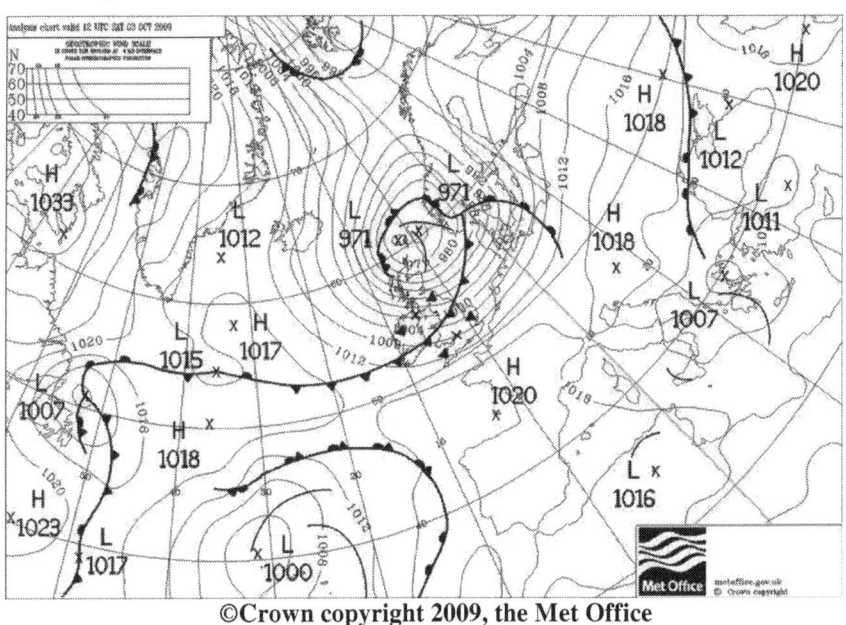

Figure 9.2 – Atlantic Chart for 12:00, 03 October 2009

©Crown copyright 2009, the Met Office

9.2 High Pressures

These are slow moving, innocuous, gentle giants. Take a look in Figure 9.1 at 53° north and 29° west, this shows the centre of a high

84

Chapter 9 – An Introduction to Weather Charts

at 1029mb. On the chart above, two days later, it has been distorted and squeezed, but it has hardly moved, compared with the frontal low, which moved 1,300 miles in the same time!

When a high is over the UK, it spells dry, settled weather and, because it is slow moving, these conditions can persist for many days or weeks.

Unfortunately, there is not too much more to say, other than to show you how a high pressure influencing our weather can bring either very good or very bad weather. The first example, from September 2009, shows mild, settled weather, and the second, from February 2009, shows very cold conditions, with heavy snow.

Below, Figures 9.3 and 9.4 are the MSL pressure chart and the Jet Stream chart for 25 September 2009. The weather throughout September 2009 was very settled, dry, and mild. (You will be pleased to know that during September my tomatoes eventually pulled through, and I got a good crop, even though it was two months late!)

Understanding the Jet Stream – Clash of the Titans

The Atlantic Chart and Jet Stream Chart for 25 September 2009

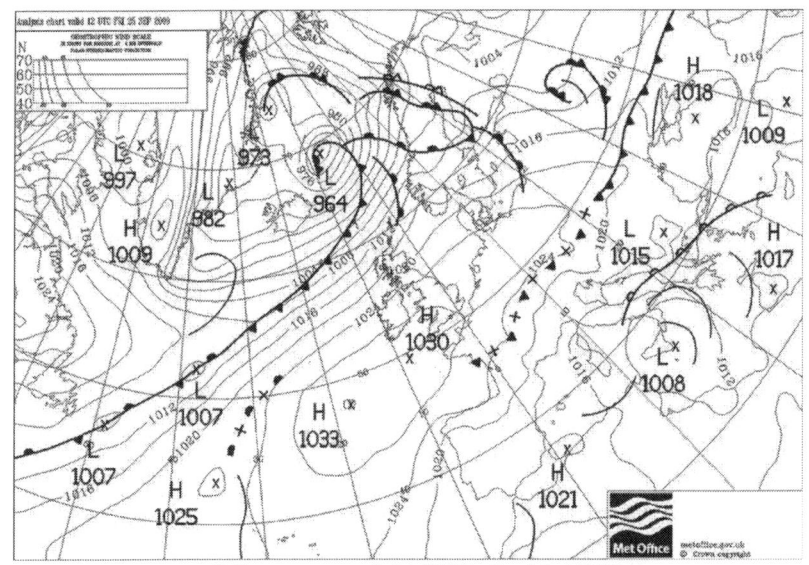

©Crown copyright 2009, the Met Office
Figure 9.3

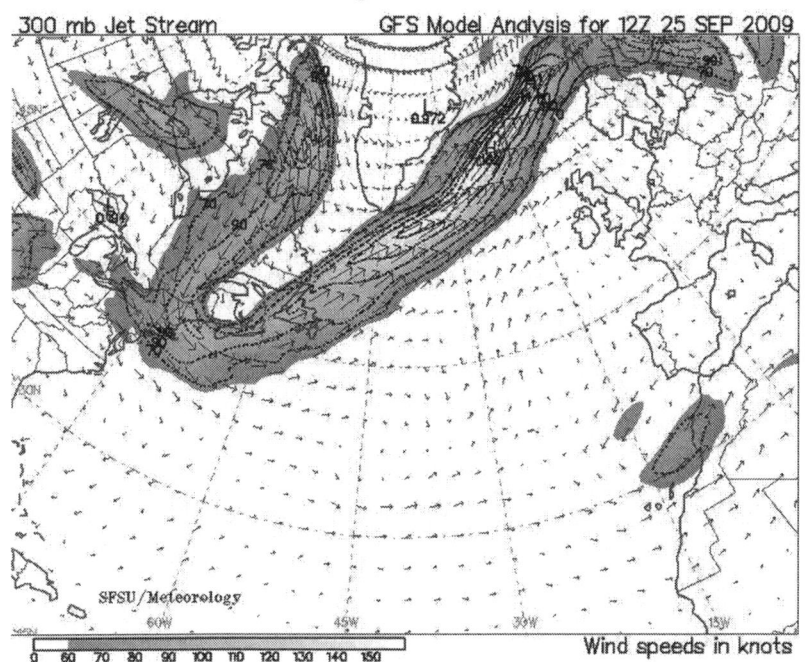

Department of Geosciences at San Francisco State University
Figure 9.4

Chapter 9 – An Introduction to Weather Charts

From the MSL pressure chart, you can see that there is a high pressure system over the UK and that the air is being drawn in from the warm latitudes of the Atlantic. The Jet Stream shows an Arch, which signals clement weather. For almost the whole of the month of September, the Jet Stream did not enter our air space. The profiles for the month were The Arch, The Horseshoe and None – all of these mean that our weather is governed by the prevailing high and low pressure systems; these, in the main, were highs from the Atlantic and near continent. Temperatures were above average and the rainfall was down to 40 percent of the average – all in all, a lovely September.

Now here's an example of a high as the bad guy:

Understanding the Jet Stream – Clash of the Titans

The Atlantic Chart and Jet Stream Chart for 01 February 2009

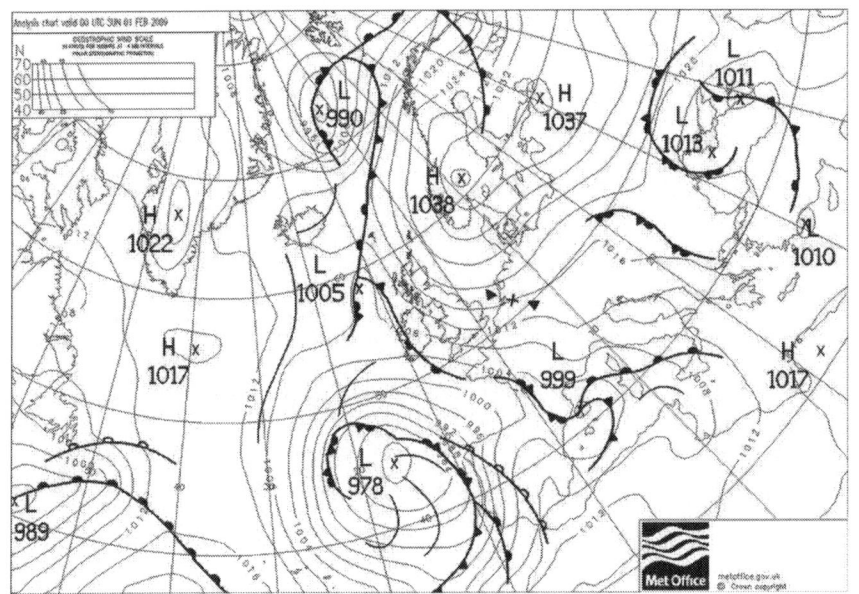

©Crown copyright 2009, the Met Office

Figure 9.5

Chapter 9 – An Introduction to Weather Charts

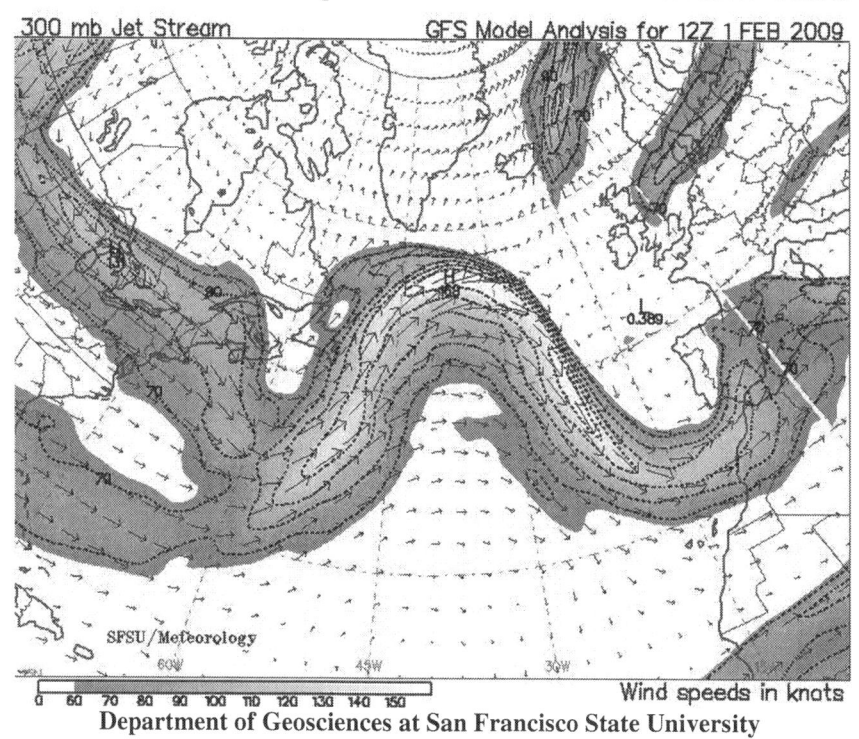

Figure 9.6

The MSL pressure chart shows us a large, complex high pressure system over Scandinavia and Western Russia; which, in February, are very cold regions. Remember, high pressure systems rotate clockwise, so the air being brought into the UK originated from the east and has passed over the cold regions of Europe. The air has had very little contact with water, and, in addition to being extremely cold, it is obviously very dry.

If you look closely, you can see that air over the southwest of England originated from the Atlantic via the low pressure off the north coast of Spain, this means warmer, moist air. When these two air flows meet, the cold air will, as usual, undercut the warmer air,

Understanding the Jet Stream – Clash of the Titans

cause condensation, and, because of the cold lower layer of air, any precipitation will fall as snow. (NB: the formation of snow is more complex than this but I think you get my gist).

The Jet Stream profile is a horseshoe (Figure 9.6), and this again illustrates that our weather is at the mercy of the prevailing high and low pressure systems.

The first week of February 2009 was bitterly cold, with moderate to heavy snowfalls, especially in the east of the country.

By complete contrast, above-average temperatures were recorded for the second half of the month (February 17 onwards). Have a look at the archived Jet Streams, and you will see that, for seven out of eleven days, the UK was affected by a series of Arch profiles. These arches had two effects: firstly they allowed high pressure in the Atlantic to build and, secondly, they formed a block on the east of the country to prevent the similar cold weather affecting us, as at the start of the month.

Chapter 9 – An Introduction to Weather Charts

9.3 The Fronts

I am hoping, since you have read Chapter 7, you are now familiar with the fronts shown on MSL pressure charts, and that I don't need to say much more. But, take a look at Figure 9.3 (page 86); there are two examples of fronts, which I have not mentioned. One stretches across Europe from 46° north, 1° west to 48° north, 20° east and the other 42° north, 14° east to 36° north, 33° east. There are six front types that I have not discussed – primarily because they are academic and not really significant to us mere mortals.

The first is a weakening cold front and the second a weakening warm front. Below, in Figure 9.7, are the symbols for these six types of fronts.

Weakening Cold Front   Weakening Warm Front

Developing Cold Front   Developing Warm Front

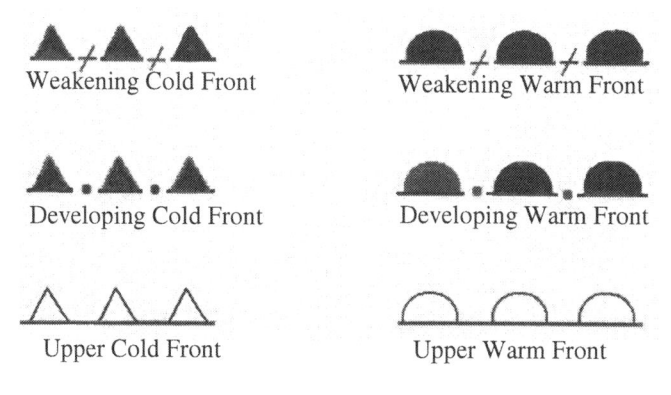

Upper Cold Front   Upper Warm Front

Figure 9.7

Further details are available on the Met Office Web site.

91

9.4 Isobars

Isobars are analogous to the contour lines on an Ordnance Survey map, which join points of equal height.

Take a look back at Figure 9.1, there is a high pressure zone, marked H, south-east of Greenland; this is surrounded by three concentric isobars. You will see on the outer isobar (south of Greenland) the number 1020, this tells us that all the points on this line have an MSL pressure of 1020mb. Off the coast of Ireland, on the middle isobar of the three, you will see the number 1024 and the inner one does not have a number. The reason why no number is shown is to prevent cluttering; do I hear you asking, what's the point, then? Well, the Met Office has a standard such that all isobars are spaced in steps of four millibars. Because the outer two isobars are increasing towards the centre (marked H), then the inner isobar must be 1028mb and the maximum pressure is at point H, at 1029mb, the analogy here is the top of a mountain on a survey map.

Notice that usually the isobars round a high are spread out, compared with a low, this is because highs normally cover a larger area than a low. So, comparing this with contour lines – take an increase of height of, say, 1,200 feet - if we were to apply this to the rolling South Downs the contour lines would be spread over quite a sizeable area, while in the Peak District, it could represent a small mountain sticking out of a ridge.

Chapter 9 – An Introduction to Weather Charts

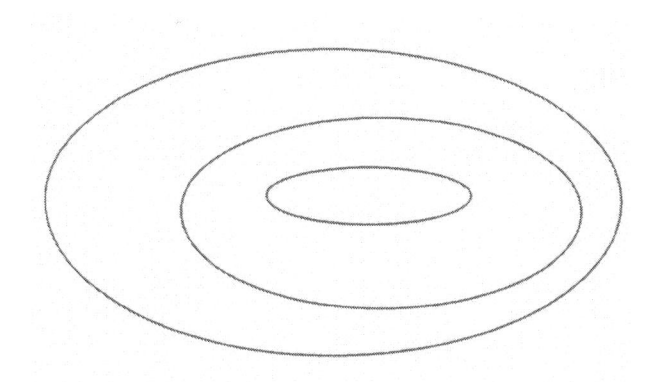

Figure 9.8 – The South Downs

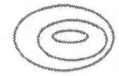

Figure 9.9 – Peak District

What else do isobars tell us? Back to the contour lines; we are going to visit both our Peak District Mountain and the South Down hill when they are covered with snow (bear with me) and climb to the top of both of them, with sledges on our backs. *Wheeeeee*, down the mountain we go, it only takes a minute to get down. We repeat this on the hill, and it takes over five minutes. It is obvious why; the slope of the mountain is greater than that of the hill!

This is the same as our highs and lows – the air exiting a high is much slower than that of the air entering the low – but in this case it is the pressure gradient and not the height gradient. To be absolutely realistic, the South Down case does represent a high pressure as the movement is way from the centre (the top of the hill), but the Peak District example should be represented as a funnel shaped chasm so that we are sliding towards the centre.

We know that air circulates around a high in a clockwise direction, so, when there is a high pressure around, then the winds are going to be light (very light) and, facing the centre of the high the winds will always blow from the right.

Now, as usual, lows are the complete opposite. Firstly, we know that they circulate anticlockwise and secondly, that the wind speed and direction are not as straightforward. Using our contour line analogy, we could have a gently sloping hill in the Peak District and the contour lines spread out similar to the downs, so the wind speed would be light. But, by contrast, we could find a near-vertical cliff face, where the contour lines would be very tightly packed – this would result in hurricane-force winds. So, the winds around a low vary, depending on the closeness of the isobars.

We know that the air circulating around a high just drifts outward and this motion has very little effect on the wind direction, but the air round a low is being sucked in. The direction of the wind associated with a low depends on the distance you are from the centre. Facing the centre of the low, a fair distance away from the centre, the wind will blow from the left, but, as you move closer to the centre, there will be a tendency for it to move behind you – all very confusing.

Chapter 9 – An Introduction to Weather Charts

But, when you're looking at a chart to figure out the wind direction, the rule of thumb is face the centre of a high, wind from the right and facing the centre of a low, wind from the left.

Well, that just about finishes this section on weather charts. I find them fascinating, looking at the Met Office's three-day forecasts, watching the development of high and low pressure systems and tracking frontal depressions. What is even more satisfying is making my own forecasts, especially when the Met Office gets it wrong!

Chapter 10 – Forecasting using JS and Atlantic Charts

In the UK, we are obsessed with our weather, and understandably so, because it plays such an important role in our daily lives. Our weather is so fickle that any pursuit that involves going outside is affected by what the weather is going to do (or not do); it affects anything from the mundane routine of going to work, to planning summer or winter holidays. So, it is important to all of us to have a rough idea of what is going to happen over the next couple of days - hence the need for weather forecasts.

I am not going to go into the history of weather forecasting other than to say that, since the advent of the space age and supercomputers, it is now *far* more advanced than even thirty years ago.

Below are some examples of why forecasting is important forewarned is forearmed, as they say!

Personal	We need to know what to wear and what accessories to carry and when not to venture out.
Sports events	For sailing, hiking, mountaineering, gliding, hot-air ballooning, etc. (In June last year we went up in a hot-air balloon and it took ten months from booking the flight to actually going up – due to the weather!)
Severe Weather	We can prepare for storms, snow, winds, heat waves, extreme cold, hurricanes, tornados, drought, floods, etc.
Travel	All forms of travel can be influenced, whether over land, sea, air, or even into space. Local, national, and international journeys are all included.

Chapter 10 – Forecasting Using JS and Atlantic Charts

Weather forecasting is not an exact science; it would be good if it were – we could all do it. The supercomputers produce short, medium and long-term forecasts for the MSL pressure, the Jet Stream, temperatures, humidity, etc. Unfortunately, these are all vulnerable to the elusive *"Chaos Theory"*, which means that a slight change in any one of the variables can have a dramatic effect on the overall outcome. For this reason, you should always take medium and long-term forecasts with a pinch of salt; it is difficult to predict with confidence more than a few days in advance, and longer-term forecasts should always come with a probability caveat. The national media are guilty of taking forecasts as gospel – the "barbeque summer" of 2009 (forecast by the Met Office in April 2009) was reported as a certainty, but in actual fact, they had advised a 67 percent chance of a good summer, so there was a 33 percent chance that it would not happen, and it didn't!

However, don't get me wrong; forecasts these days are far more accurate than even in the 1970s. For instance, the MSL pressure forecasts - both the one and two day forecasts - are now 75 percent more accurate, and the three day 64 percent improved (calculated using the differences between the actual and predicted pressures).

Before we get to the nitty-gritty of forecasting, I need to briefly mention a couple of points of interest, which go hand in hand: electromagnetic radiation and satellite imagery. These just add the icing on the cake to weather forecasting.

10.1 Electromagnetic Radiation (EMR)

This sounds horrendous, doesn't it? It is one of those scientific mouthfuls that cannot possibly have any relevance in every-day life, but you'll be surprised. So, what is it then? Have you got a microwave oven? Have you had an X-ray recently or listened to the radio? You are reading this book – the reason you can see the words is because you have special sensors that are sensitive to a certain type of electromagnetic radiation, called light, and the sensors are your eyes. A microwave oven uses another type of Electromagnetic Radiation (EMR), which excites the water molecules in the food, giving them energy, which causes the food to heat up and cook. Both X-rays and radio waves are again different types of EMR. The different types are the result of the EM radiation vibrating at different frequencies; this is analogous to sound - the pitch of the sound changes with a change of frequency; the faster the air vibrates, the higher the pitch of the sound we hear.

The examples above form part of what is called the EMR spectrum – most of us are aware of the visible light spectrum, violet through to red, as displayed by a rainbow, but visible light only forms a very small fraction of the whole EMR spectrum. At the top end (having the highest frequency) are gamma rays (radioactive decay); then going down the spectrum we have: X-rays, ultraviolet radiation (which causes sunburn), the visible spectrum, infrared (this one is important – see below!), microwaves and last, the snail, radio waves. I am not going to bore you with actual frequencies, I'll just say that gamma rays vibrate 10^{11} times faster per second than radio waves!

Chapter10 – Forecasting Using JS and Atlantic Charts

The relevance of all this is that the sun bombards us with a wide variety of EMR. Luckily for us, our atmosphere absorbs a lot of it, although, thankfully, not light (the ozone layer is responsible for absorbing the dangerous ultraviolet light). In this basket of EMR, there is one of particular interest – infrared (which means "below red", and this is where it is in the EMR spectrum). You may have learnt at school the three methods of heat transference: convection, conduction, and radiation – the radiation method was always the difficult one to visualise. This is because it is EMR at a slightly lower frequency than red light – it is heat transfer by infrared radiation. All objects that are warmer than their surroundings emit infrared EMR, and as a general rule, the warmer they are the more radiation they emit.

Let's put this in context of the weather – the sun warms the earth's surface (by radiation), which makes it warmer than the surrounding air and space – so the earth will emit infrared radiation (the air that is actually in contact with the surface will also be heated by conduction, and the air above that by convection!). Well, if the earth is radiating infrared light back into space (not by reflection but by emission), it should be possible to detect this radiation. This is exactly what some of the weather satellites do. What's the point? Surely the images would all be identical. Well, no, they're not, because it is a measure of "temperature" radiation, either from the earth's surface (land and oceans) or the tops of clouds. Temperature decreases with altitude, so the higher the cloud top, the cooler it is, and, hence, a lesser amount of infrared (IR) radiation is emitted into space.

The satellite cameras are sensitive to IR, and, the stronger the radiation, the darker the image – like a negative image from the old days of black and white photographic film. As a guide, black means no cloud (radiation direct from the earth's surface), dark grey means low cloud tops, and light grey means high cloud tops.

One distinct advantage of IR images is that they work perfectly well during the night – the earth and the oceans do not switch off the IR emissions just because it is dark. So, let's have a look.

10.2 Satellite Imagery

All of the previous charts have been forecasts, but we now have access to what is actually happening, not quite in ***real time***, but near enough – these are satellite images covering the entire earth's surface.

There are a lot of these meteorological satellites all with different orbits. Some are in geostationary orbits (i.e., they remain fixed over one point of the earth's surface), and some gallivant all over the place. The pictures they take are high quality and usually in black and white, and they are often computer enhanced to show colour.

I use the site www.eurometeo.com/english/meteosat, which has the following images available:

Visible Images – These are just pictures taken of the earth as if you were watching from the satellite (remember light is EMR, but in this case, reflected). Updated every hour on the hour.

Chapter10 – Forecasting Using JS and Atlantic Charts

Infrared Images – We know what these are now. Updated every hour on the hour.

Water Vapour (WV) Images – These are special IR images that are created by analysing, using onboard computers, the IR that has passed through our atmosphere, any substance that allows EMR pass through is known as a medium (air, water, glass etc). After any EMR has passed through a medium, it displays a certain characteristics; these are unique to that combination of EMR and the medium (known as absorption lines); this property of air changes depending on the amount of water vapour it contains. So, if we know the base characteristics for water and air, we can compare the properties of the IR light received from the earth to this base and then estimate the WV content of the air that the IR has passed through. The images are similar to the IR images – dark meaning very little water vapour to nearly white meaning a high water vapour content. Again, these are updated every hour on the hour.

Precipitation Estimates – These are a natural progression from the Water Vapour Images and, again, they use a bit of computer jiggery-pokery to display the estimated precipitation, in millimetres per hour, that the water vapour images are likely to produce. They are colour-coded (light blue, 1 mm/hour, through to red, 35 mm/hour). They are updated every hour on the hour.

These are "good fun" – you can follow a band of rain moving across the country and zoom in when it gets close to you. Then you can phone a friend (don't ask the audience!) and predict with confidence that they are going to get some rain at such-and-such a time.

Understanding the Jet Stream – Clash of the Titans

To view the UK and Europe select Western Europe. Below, Figures 10.1, 10.2, and 10.3 are examples taken from today, 01 November 2009, all at 00:00 UTC. A frontal low was moving in from the Atlantic, and the forecast was heavy rain and strong winds across the entire UK.

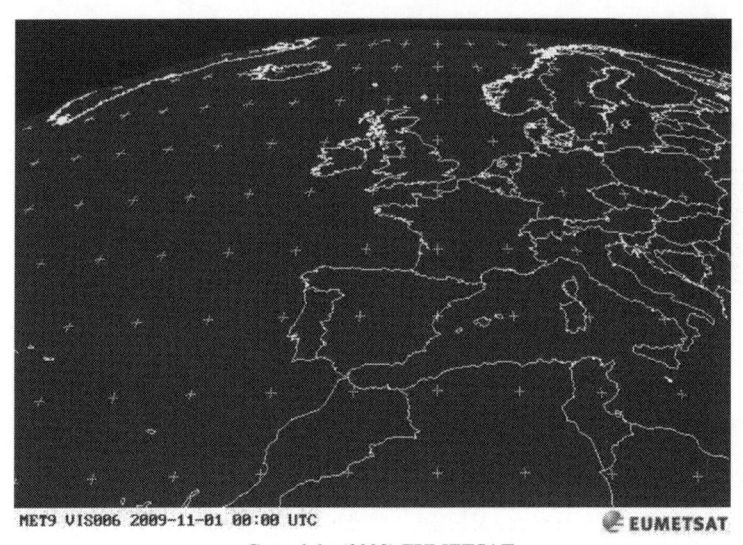

Copyright (2009) EUMETSAT
Figure 10.1 Visible Image

Chapter10 – Forecasting Using JS and Atlantic Charts

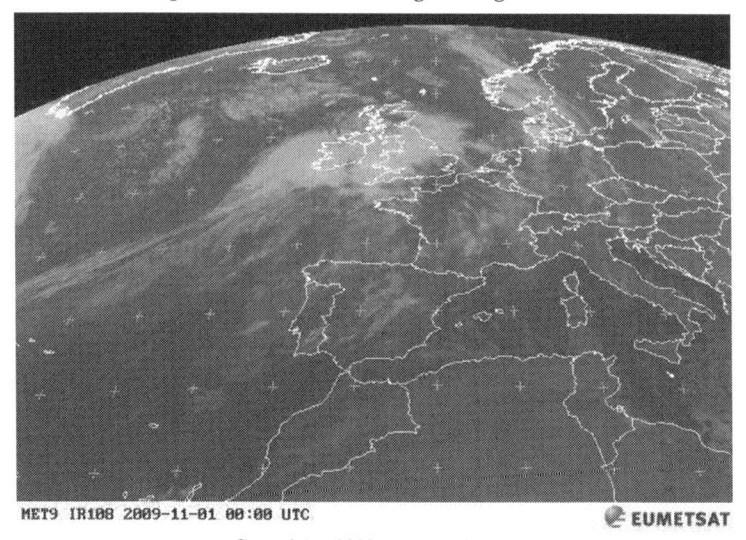

Copyright (2009) EUMETSAT
Figure 10.2 Infrared Image

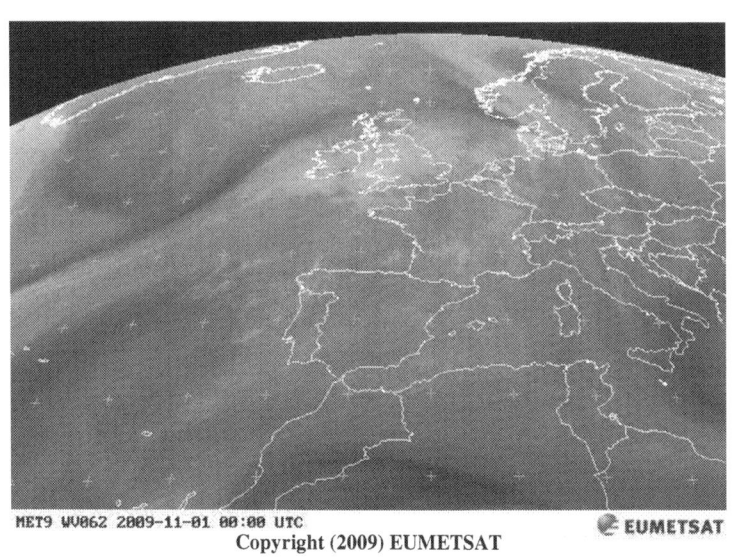

Copyright (2009) EUMETSAT
Figure 10.3 Water Vapour

This is a really good illustration of the power of IR images. These images were taken at 00:00 in the morning and, obviously, the visible image shows nothing, as it is midnight, whereas the IR image still shows us a cloud distribution. The country outlines and the "+" marks are superimposed. The "+" marks – these are points where longitude (in steps of 5°) and latitude (steps of 5°) cross. To make any sense, there has to be a base point, for instance the + in the middle of Spain is 40° north and 5° west. These reference marks are useful when calculating approximate speeds of weather features (low pressure system, rain-bearing cloud, etc.). Please remember to use **spherical trigonometry** when estimating distances using longitude and latitude.

Figure 10.2 (IR) shows a swathe of cloud over Scandinavia and more importantly, a huge cloud mass over most of the UK.

There is cause for some confusion here; the IR image shows the large cloud mass as light grey, which implies high cloud tops. For example, the tops of high *cirrus* clouds are cold so they will emit very little IR, but the top of a tall *cumulonimbus* cloud can also be very high, therefore cold, and will also emit little IR. So, how do we know which is which? Just take a look at the water vapour image, Figure 10.3. Our two cloud masses are both very light grey meaning a high WV content, and they are probably rain-bearing clouds. During the morning of 1 November 2009, most parts of the UK were soaked, with some regions getting in excess of 50mm of rain.

10.3 Forecasting

Practice makes perfect – remember that all aids to forecasting (Jet Stream, Atlantic Charts, IR, and air water vapour content satellite images) can change, so it is a continual process.

Just a reminder: the faster the Jet Stream speed, the greater effect it can have on the surface winds – but the direction and speed will always be dependent on the distribution of the isobars. What can happen, though, is that the MSL pressure forecasts change (over a couple of days) as the influence of the Jet Stream takes effect.

If you've managed to get as far as reading this, it probably means you are, like myself, fascinated by the Jet Stream and are interested in having a go at weather forecasting. So, below is a simple flow chart showing a rough guide to forecasting; as I have said before, it is not an exact science, and practice makes perfect.

You will see on the second page of the flow chart (label 5) that I advise you to "Check MSL for PHL, fronts, isobars etc", so below I have included quick reference tables summarizing the conditions of the passage of different front types and a frontal low.

10.3a Cold Front

Cold Front	Approach	Passage of the Front	Comments
Tempe		Sudden Fall	
Wind Speed	Light	Rises	
Cloud	High cirrus an hour before the front arrives, then low cloud prior to the arrival	Clouds thicken; chance of cumulonimbus	Cirrus cloud will only appear if there is a fairly strong upper wind
Rain	No	Heavy	Cloud and precipitation dependent on the state of the warm air: Moist and stable = rain unlikely Moist and unstable = rain likely Dry = rain unlikely
Duration	± 1 hour		

10.3b Warm Front

Warm Front	Approach	Passage of the Front	Comments
Temp	Steady	Rises	
Wind Speed	Light	Variable	
Cloud	Variety of "gentle" clouds: stratus, cirrus or perhaps nimbostratus	Cotton wool - stratocumulus	Clouds may appear well ahead of the front (up to 1200km), but by the time the front arrives (±2 days) you will probably have forgotten!
Rain	Possible – light and steady immediately ahead of the front	Very little	
Duration	± 2½ days		

10.3c Occlusions

It is unlikely that you will experience the formation of an occlusion, as they are usually well established before they reach us. But, the passage of a developed occlusion over the UK is quite common and that means gusty winds and heavy rain (not very nice at all). They are usually over and done with in a couple of hours.

Chapter 10 – Forecasting Using JS and Atlantic Charts

10.3d Passage of a Frontal Low

So, what can we expect when a frontal low lurks over the western horizon? Watch out (I could end the section here! But I won't).

Below is Stage 5, rotated through 90°, from section 7.5, a typical frontal low motion.

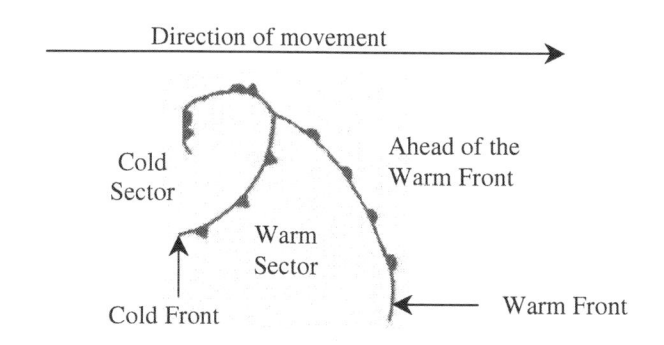

	Ahead of Warm front	Passage of Warm Front	Warm Sector	Passage of Cold Front	Cold Sector
Temp	Cool begins to rise	Continues to rise	Mild	Dramatic drop	Stays cold
Wind	Increases	Blustery, strong gusts	Steady	Increases, chance of gale force	Squally
Cloud	Cirrus then altostratus	Low nimbo-stratus	Thins, perhaps patchy	Thickens chance of cumulo-nimbus	Thins, culumus
Rain	Yes, the nearer to the front	Yes, can be heavy	Becomes lighter	Heavy, perhaps hail/sleet	Showers
Duration	1 to 2 days, may get "stuck" and drop many inches of rain – e.g. Cockermouth, Cumbria, November 2009					

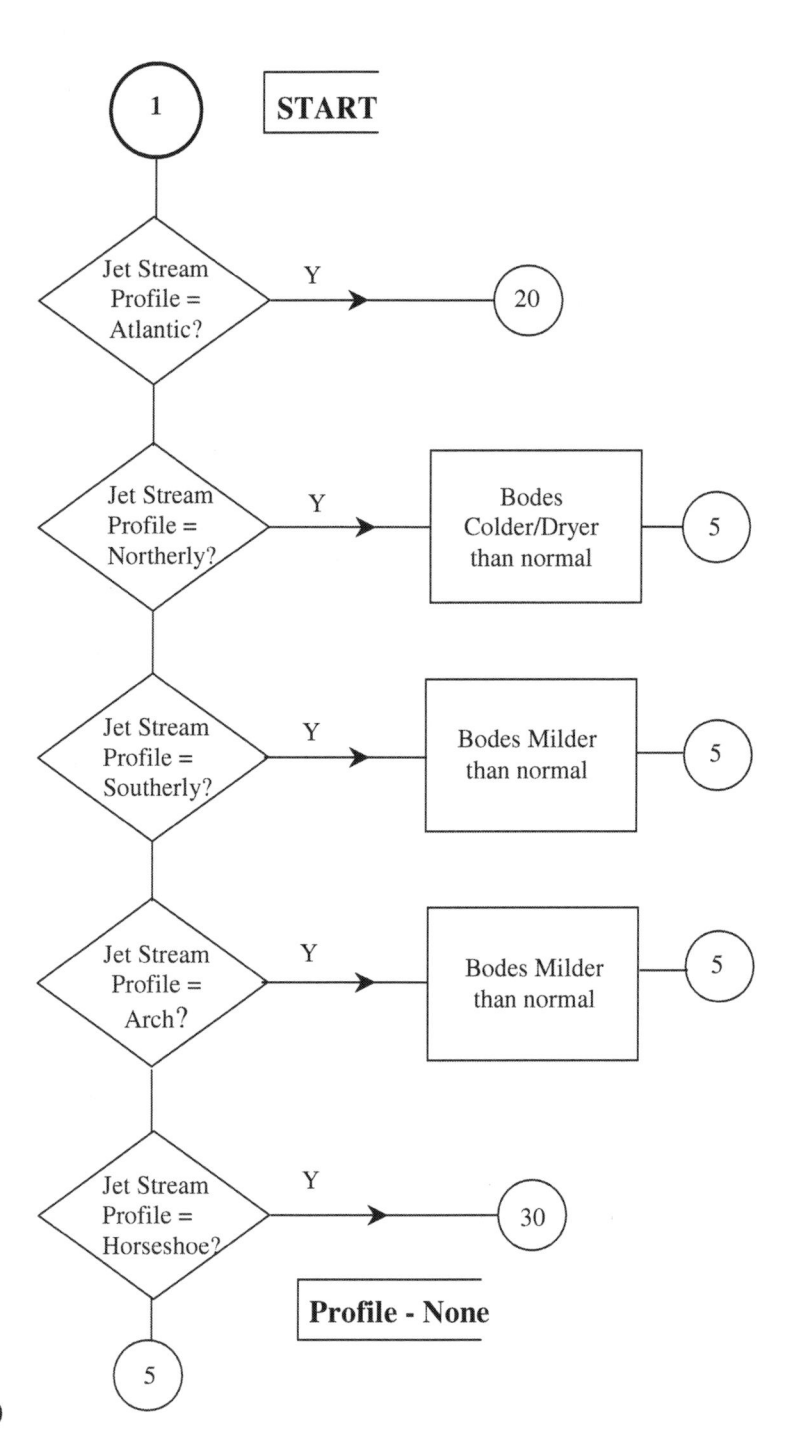

Understanding the Jet Stream – Clash of the Titans

```
         ( 1 )      [ START ]
            |
            v
      /Jet Stream\      Y
      < Profile =  >-------> ( 20 )
      \ Atlantic? /
            | 
            v
      /Jet Stream\      Y    ┌──────────────┐
      < Profile =  >-------> │    Bodes     │──( 5 )
      \ Northerly?/          │ Colder/Dryer │
            |                │ than normal  │
            v                └──────────────┘
      /Jet Stream\      Y    ┌──────────────┐
      < Profile =  >-------> │ Bodes Milder │──( 5 )
      \ Southerly?/          │ than normal  │
            |                └──────────────┘
            v
      /Jet Stream\      Y    ┌──────────────┐
      < Profile =  >-------> │ Bodes Milder │──( 5 )
      \   Arch?   /          │ than normal  │
            |                └──────────────┘
            v
      /Jet Stream\      Y
      < Profile =  >-------> ( 30 )
      \ Horseshoe?/
            |           [ Profile - None ]
            v
          ( 5 )
```

110

Chapter10 – Forecasting Using JS and Atlantic Charts

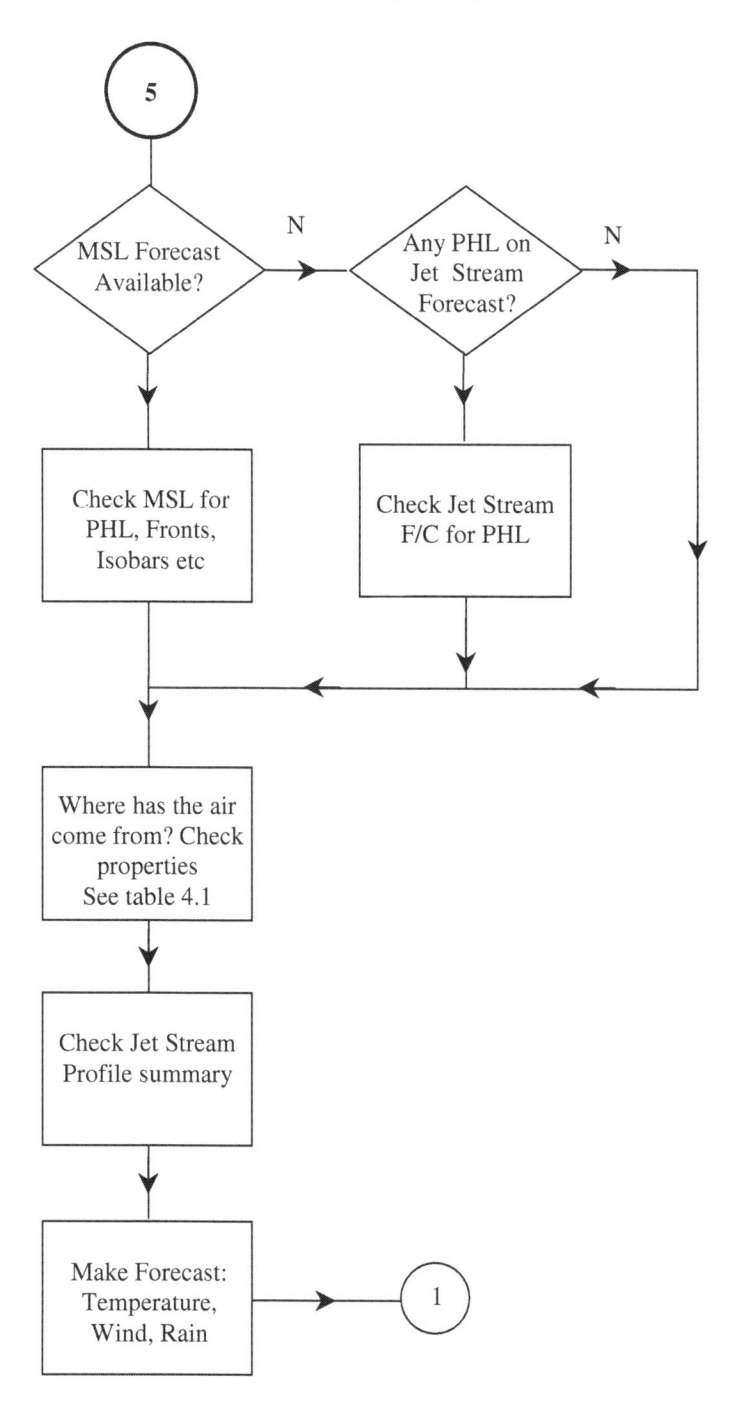

```
                    ( 5 )
                      │
                      ▼
              ┌───────────────┐       N        ┌───────────────┐    N
              │ MSL Forecast  │──────────▶    │  Any PHL on   │──────────▶
              │  Available?   │                │  Jet Stream   │            │
              └───────┬───────┘                │   Forecast?   │            │
                      │                        └───────┬───────┘            │
                      │                                │                    │
                      ▼                                ▼                    │
           ┌────────────────────┐          ┌────────────────────┐           │
           │   Check MSL for    │          │  Check Jet Stream  │           │
           │    PHL, Fronts,    │          │     F/C for PHL    │           │
           │    Isobars etc     │          │                    │           │
           └──────────┬─────────┘          └──────────┬─────────┘           │
                      │                               │                    │
                      ◀───────────────────────────────┴────────────────────┘
                      │
                      ▼
           ┌────────────────────┐
           │ Where has the air  │
           │ come from? Check   │
           │     properties     │
           │   See table 4.1    │
           └──────────┬─────────┘
                      │
                      ▼
           ┌────────────────────┐
           │  Check Jet Stream  │
           │  Profile summary   │
           └──────────┬─────────┘
                      │
                      ▼
           ┌────────────────────┐
           │   Make Forecast:   │
           │    Temperature,    │────────▶ ( 1 )
           │    Wind, Rain      │
           └────────────────────┘
```

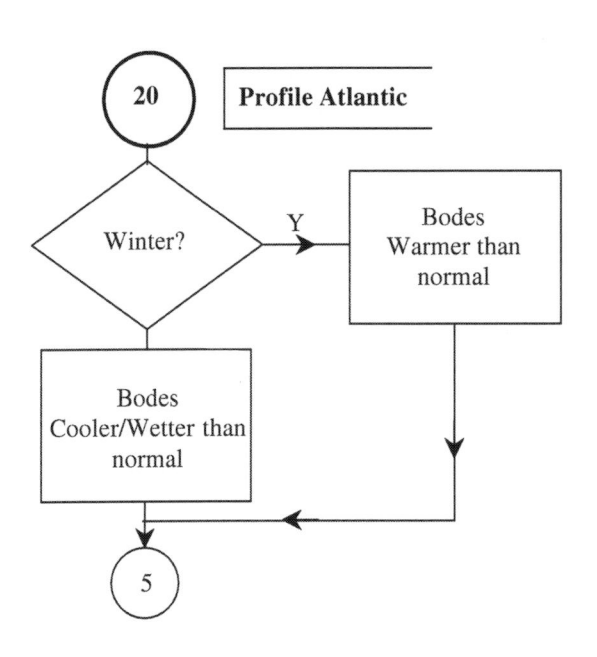

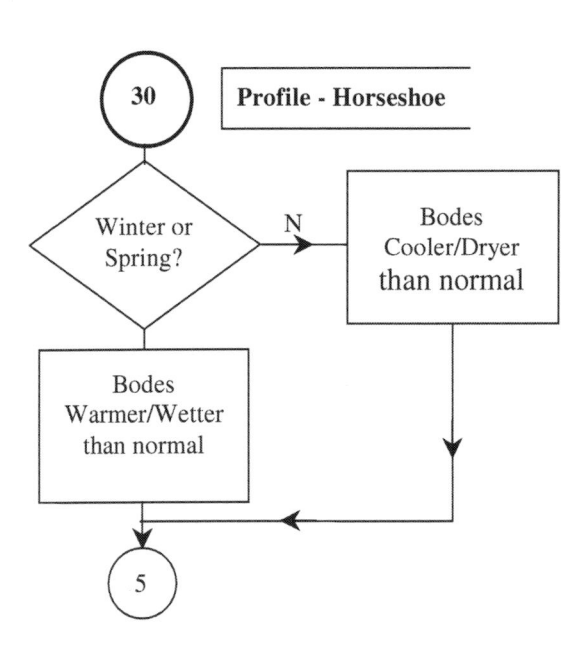

Understanding the Jet Stream – Clash of the Titans

20 — **Profile Atlantic**

- Winter?
  - Y → Bodes Warmer than normal
  - (N) → Bodes Cooler/Wetter than normal
- → **5**

30 — **Profile - Horseshoe**

- Winter or Spring?
  - N → Bodes Cooler/Dryer than normal
  - (Y) → Bodes Warmer/Wetter than normal
- → **5**

Chapter 10 – Forecasting Using JS and Atlantic Charts

Key to Flow Chart Symbols:

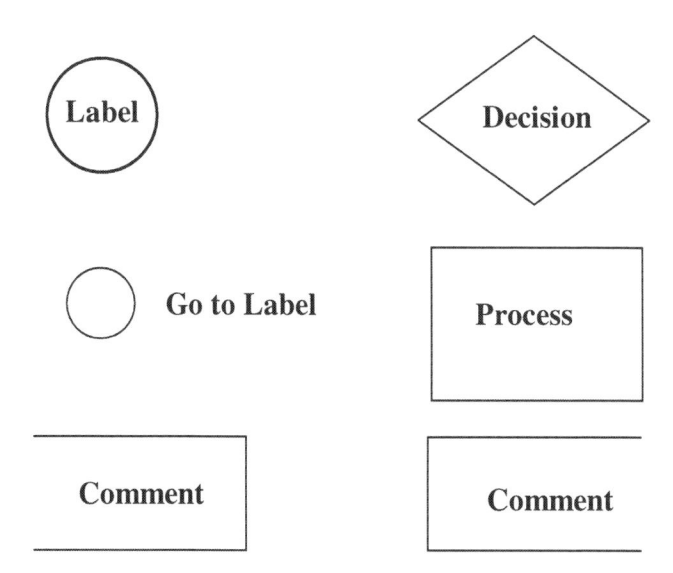

(Label)	◇ Decision ◇
○ Go to Label	[Process]
[Comment]	[Comment]

Understanding the Jet Stream – Clash of the Titans

When I do a weather forecast I use three sources of information:

1) Jet Stream forecast
2) MSL Atlantic chart
3) West Europe and West Atlantic Satellite images for WV and infrared data.

You cannot use the Jet Stream in isolation but valuably it gives the sort of air that will be present (17/18/19/20 August 2008 forecasts)

Something you will have noticed is that I have only mentioned winter and summer, so what about spring and autumn? These are the transitional months (rather obviously). The only thing I can say, really, is practice and take particular notice of the source of the air.

When you see a weather systems approaching from the Atlantic, ballpark figures for the time they take to reach us would be:

From the Newfoundland region ±3 days

From the Greenland region ±2 days

From the Iceland region ±1 day.

Chapter 11 – The Endish

11.1 – Practice Makes Perfect

Now that you have reached here, perhaps you should have a go at your own forecast, no, seriously, it's time! But don't make it too easy, have a go at a four-day forecast using only the Jet Stream charts (the Met Office only give a three day Atlantic Chart forecast). Go to www.squall.sfsu.edu/crws/jetstream.html. The right-hand column is North Atlantic; scroll down and you will see Forecasts, Click Big Images; on the pop-up screen, Click Build Animation – this gives twelve hour forecasts for the next 4 days. Study the last two images, and look for:

- What type of profile it is?
- Is it over the UK or your town?
- Is it a powerful jet?
- Where is the air coming from?
- Are there any obvious low (near zero knots) jet speeds?
- If so, are the surrounding jet winds moving clockwise or anticlockwise? This will indicate the position of any obvious high or low pressure systems.

Now, put your deerstalker hat on, get out the magnifying glass, be bold, and make an elementary (dear Watson) prediction – temperature, precipitation, and general wind strength and direction. Write it down and tweak it as the Atlantic charts become available plus check again the Jet Stream forecasts. This is nothing to be ashamed of it's exactly what the professional forecasters do. Then, on the actual day – see how close you were to your original forecast! You may be surprised. *Good Luck.*

11.2 – Can You See the Jet Stream?

No, not directly. However, it is possible under the right conditions, to see a direct influence.

On a clear day, if you notice a high cirrus cloud that looks like an abnormally wide aeroplane vapour trail (contrail), and you can actually detect slight movement, check the position of the Jet Stream. If it is above you and moving in the direction of the cloud, then the cloud is probably being driven by the Jet Stream.

Believe or not, this is not a rare event - in the last four months I have witnessed it three times.

11.3 - The Jet Stream – Can Be Quite Nasty

It is difficult to detail past events that were caused directly by the Jet Stream, primarily since its existence has only been known for the last sixty odd years, hence there is no archive material. But, rather than bore you with a lot of detail I will mention four particular events, which can be checked out, on the Net.

1. The 1947 disappearance of the "Stardust" aircraft over the Andes – this one is particularly fascinating, because this passenger plane disappeared without a trace, and all sorts of theories emerged (aliens, a new Bermuda Triangle, etc) -

Chapter 11 - The Endish

have I got your interest? Well, in that case, I will leave it there for you.

2. The Great Storm, 15/16 October 1987 (UK). I was out of the country so can't give you a first-hand account. But in France and Southern England it uprooted six million trees. Fatalities were few, as it made landfall late at night.

3. The infamous Tornado Alley (USA), an occurrence between March and June when over 1000 tornados are reported annually. Tornados are one of the most destructive weather features, but, luckily, they are usually short-lived and the destruction generally confined to a narrow band.

4. The 1930s USA Dust Bowl. For nearly a decade the southern plains of the USA were devastated by a prolonged severe drought, resulting in unprecedented dust storms. The main cause was man-made, due to extensive farming and ignoring agriculture's golden rule of crop rotation. However, it is believed that the Jet Stream had a part in the creation and sustaining of the drought.

11.4 - The Jet Stream and Climate Change
The first statement to make is that any unseasonable weather is the result of the Jet Stream not being where it should be at that time of year. The corollary to the above is that any seasonable weather is a result of the Jet Stream being where it should be!

The normal cycle is considered to be, in summer, the Jet Stream should be north of the UK; then, through autumn, it moves south,

Understanding the Jet Stream – Clash of the Titans

until it is below the UK in winter, and then through spring, it moves back northwards.

For instance, the summer and autumn of 2009 were classic examples; July was the wettest on record, because the Jet Stream spent a lot of the month as an Atlantic profile (across us), bringing an unseasonable July. But September and October were exceptionally mild and dry – the Jet Stream was above us, in the Arch profile.

But what will happen if the climate changes? I think most of us are aware of the three schools of thought on the reason for climate change and global warming:

1. It is solely the responsibility of human activity.
2. It is a naturally occurring phenomenon caused by solar and terrestrial cycles over thousands of years, and there is nothing we can do about it.
3. It is a combination of points 1 and 2 – and we are only exacerbating the problem.

I will give you my opinion (which doesn't count for anything – poor Ged). I believe it is almost certainly a combination of natural and our activity – my thoughts are:

The world's land-based fauna have been flourishing on earth for millennia, and our contribution is small compared with the belching, flatulence, and any other excretions from the animals, insects, etc. Also, volcanic activity is responsible for pumping thousands of tons of carbon dioxide, sulphur, and other pollutants into the atmosphere.

Chapter 11 - The Endish

In the past, volcanoes were much more active than they are now – yet the natural cycles of warming/cooling continued – so there seems to be a certain inevitability - no matter what we do, the climate is going to warm over the next 100 to 200 years.

How would a global temperature increase affect the Jet Stream? We have to think back to basics – the Jet Stream can only be formed when the temperature gradients of the air masses are above a critical value. So, one can assume that with climate change the temperature of the air masses (polar and subtropical) would normalise and that the Jet Stream would at least weaken or even disappear (as it can during the summer). Our weather would then be at the mercy of the near continent – wet, hot summers and cold, dry winters.

We do have to factor in other considerations – how is climate change going to affect another of our weather's major driving forces, the Gulf Stream? If it disappeared, then our temperate climate would completely vanish – again we'd have hot summers and very cold winters. Most of us are aware of the dramatic seasonal weather and temperatures of New York – yet surprisingly, London is ten degrees further north - it is the Gulf Stream that keeps us warmer in winter and cooler in summer than New York.

What will be the consequences of prolonged global warming?

- ➢ The world's population is growing geometrically, so food production and water availability will have to be a priority – if not, then it could quite easily lead to worldwide warfare.
- ➢ Disease will be rife because of limited water resources.

Understanding the Jet Stream – Clash of the Titans

- The proposed new energy sources - nuclear, solar, wind, tidal etc., are not as efficient as forecast, which will probably lead to rationing.
- Energy bills and taxation will inevitably rise.
- Entire countries may become bankrupt.
- Current planning is proposing a massive expansion of new housing on flood plains and other designated high-risk areas – which we suspect will become increasingly susceptible as a result of climate change.
- Fossil fuels, especially oil, will run out, unless unlimited monies can be found to explore and retrieve the oil from the inaccessible regions – the poles and deep sea. Mind you, predictions state that the polar region may become accessible – maybe a godsend?
- Rising sea levels – the classic disaster-movie theme. A lot of the world's major cities are built on or near the coast, a frightening thought.

If one assumes that climate change is going to happen, the politicians should not only ponder the ways of cutting our emissions – causing, unnecessarily, the world population's standard of living to decline (whether in first-world or third-world countries). They should put considerable effort into learning to utilise the effects of such changes. It may be just me, but rising temperatures means energy – wind, heat, and ocean power - surely these could be harnessed for the good of the world.

Chapter 11 - The Endish

Aren't I the harbinger of doom? Let us hope I'm wrong! Anyway, it's all food for thought!

This is not the absolute end as the remaining chapters are for reference.

Chapter 12 – Basic Cloud Types

Below is a table describing the basic cloud types, their height, and what type of weather is associated with them. Most of them can be subdivided, but this is not the place.

It is amazing what clouds can tell you about the current and imminent weather, even one or two days into the future. It's a really fascinating subject. (My wife thinks I'm a bit of an anorak when it comes to clouds!)

There are excellent books available, and two I would recommend are:

The Cloud Book by Richard Hamblyn –

ISBN 13.978-0-7153-2808-8, published by D&C in association with the Met Office.

Absolutely chocker with excellent colour photos and packed with informative narrative.

The Cloudspotter's Guide, by Gavin Pretor-Pinney –

ISBN 978-0-340-89590-0, published by Hodder and Stoughton Ltd

This is more of a story, and is easy and interesting reading.

Don't blame me if you get hooked, though.

Chapter 12 – Basic Cloud Types

	Base Height (Km)	Description	Rain
Low	H<2		
Cumulus		Cotton wool, nice clouds	N
Cumulonimbus		The Giants (Titans!)	H
Stratus		Miserable blanket	D
Stratocumulus		Ominous and threatening	L
Medium	2<=H<6		
Altocumulus		Mini cotton wool	U
Altostratus		Grey, covering most of the sky	L
Nimbostratus		Another dull, grey blanket	HP
High	6<=H		
Cirrus		High, wispy feathers	N
Cirrocumulus		High and speckled	N
Cirrostratus		High, thin layers	N
Fallstreak Hole a rare cirro uncinus		The aliens have arrived	N

Table 12.1 - Basic Cloud Types – what to expect

H < 2 – H is less than 2
2 <= H < 6 - H greater or equal to 2 and H less than 6
6 < H – H greater than 6
Rain - means any precipitation:

    H   Heavy
    HP   Heavy and persistent
    D   Drizzle
    N   None
    L   Light
    U   Unlikely

Chapter 13 – Useful Web Sites

Satellite images

http://www.eumetsat.int/Home/Main/Image_Gallery/index.htm

Jet Stream – Current, Forecast, and Archive

www.squall.sfsu.edu/crws/jetstream.html

www.squall.sfsu.edu/crws.html

The Met Office

www.metoffice.gov.uk

Archived Black and White Met Office MSL charts-

www.weather.hud.org.uk/charts.php

Alternative Jet Stream and Local Weather Forecasts

www.metcheck.com/V40/UK/FREE/jetstream.asp

The BBC Weather pages

http://news.bbc.co.uk/weather/

Chapter 13 – Useful Web Sites

And last but not least mine

www.understanding-the-jet-stream.co.uk

Will be up and running early March 2010. I will attempt to put "theory into practice" and do a forecast, once a week, starting with the Jet Stream alone then "tweak" it using MSL charts and revised Jet Stream charts. Plus loads more interesting stuff.

Chapter 14 – Glossary

Air Packet	A theoretical concept to aid the understanding of the mechanics of air movement up, down, and horizontally in the atmosphere. It is thought of as a infinitely small, completely pliable and insulated bubble of air.
Convection	If air (or any fluid) becomes warmer than its surroundings then it will tend to rise.
Convective	Caused by convection
Cyclogenesis	The development and life cycle of cyclones
Cyclone/Depression	An area of low pressure.
Extratropical	Regions outside the tropics
Global	A distance in excess of 2000 kilometres
Hurricane	Also known as a tropical cyclone or typhoon. This is deep, powerful depression fuelled by the very warm seas of the tropics.
Isobar	These lines join up points of equal pressure. On the Met Office MSL charts they are stepped in 4mb increments.
Knot	One equatorial nautical mile per hour, approximately 1.15 statute miles per hour
Mid-Latitudes	Two regions between 30° north and 60° north, and 30° south and 60° south
Nautical Mile	One minute of latitude measured at the equator and equal to 1852 metres (2025.37 yards). When I first checked this definition, I thought there was an error in the dictionary, because I'd always assumed it was one minute of equatorial longitude – but it is latitude.
Precipitation -	Rain, drizzle, snow, sleet, hail, etc
Real Time	Happening now
Sinusoidal	Of the sin function $y = \text{Sin}(x)$
SI Units	An international standard, generally called the metric system

Chapter 14 – Glossary

Spherical Trigonometry	A branch of mathematics dealing with angles and distances on the surface of a sphere, similar to Pythagoras's theorems on a two dimensional plane
Subtropics	Two regions between 20° north and 35° north, and 20° south and 35° south
Synoptic	A distance between 250km and 2000km
Tropics	A region around the equator between the latitudes 23.4° north and 23.4° south.
UTC	Coordinated Universal Time. This is time measured against standardised atomic clocks, but, for everyday purposes it is Greenwich Mean Time.
Westerly Wind	Wind directions are stated as from the direction they are blowing – hence a westerly wind blows from west to east

Chapter 15 – Acknowledgements

All MSL Pressure Charts by kind permission of the Meteorological Office UK - www.metoffice.gov.uk

All Jet Stream charts appear by kind permission of Dr. Dempsey and the Department of Geosciences at San Francisco State University - www.squall.sfsu.edu/crws/jetstream.html

Figures 3.1 and 3.2 by kind permission of NOAA's National Weather Service - www.srh.noaa.gov

All Satellite images by kind permission of EUMETSAT – http://www.eumetsat.int/Home/Main/Image_Gallery/index.htm

Many thanks to the following for their patience and encouragement:-
- Audrey Dunkel (mum)
- Ruth Dunkel (better half?)
- Rachael Broddle (daughter)
- Sandra Porter
- Greg Hayes
- David Frost

Chapter 16 – The Front Cover Explained

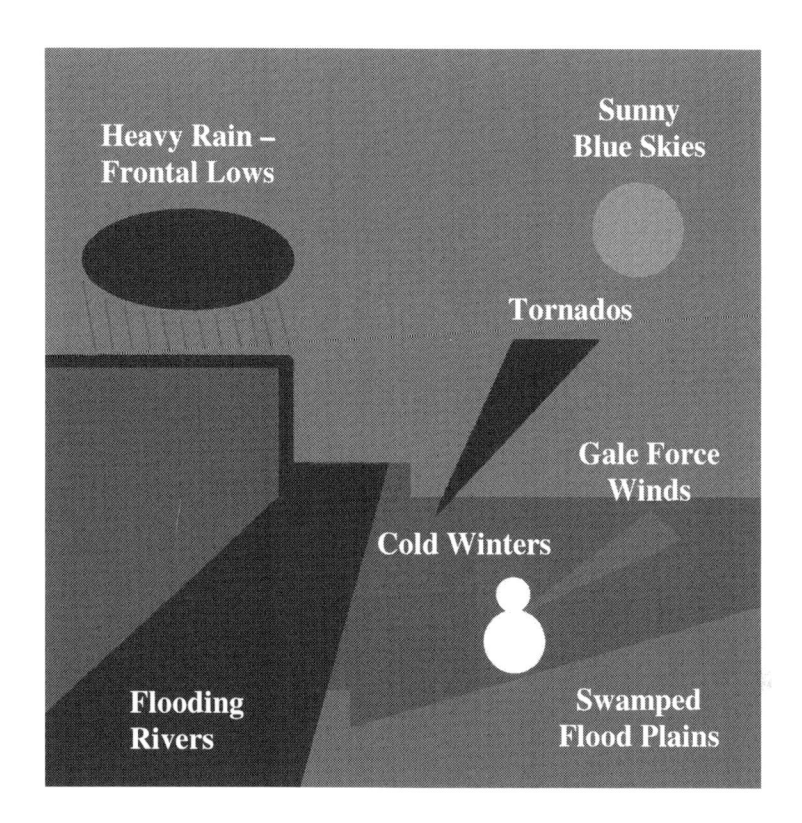

Some of the effects of the Jet Stream – clear as mud eh?

129

Printed in Great Britain
by Amazon